A REVIEW OF ALLIUM Section ALLIUM

A REVIEW OF ALLIUM SECT. ALLIUM

Brian Mathew
ROYAL BOTANIC GARDENS KEW

With advice and assistance from
W. T .Stearn (Kew) + **N. Özhatay** (Istanbul) + **J. Cowley** (Kew)

and including
Chromosome counts by **N. Özhatay** (Istanbul) & **M. A. T. Johnson** (RBG Kew)
Biochemical studies by **J. B. Harborne** & **C. A. Williams** (Reading Univ.)
Notes on leaf anatomy by **M. Gregory** (RBG Kew)

Funded by
The International Board for Plant Genetic Resources
IBPGR

First published 1996

Cover design by Media Resources, RBG, Kew

Typeset by Computype, Horton Parade, Horton Road, West Drayton,
Middlesex UB7 8EP.

ISBN 0 947643 93 1

Printed and bound in Great Britain by Whitstable Litho Ltd., Whitstable, Kent.

Prason. The Headed Leek… causes troublesome dreams.

Dioscorides, c.77 A.D.

So does its taxonomy.

B. Mathew, 1993

CONTENTS

ACKNOWLEDGEMENTS

Firstly I must acknowledge on behalf of the Royal Botanic Gardens, Kew, the work of the IBPGR and its financial backing for this project.

In addition to those colleagues mentioned on the title page I also offer my grateful thanks to the many people who have provided information or have provided living material, much of it of wild origin, notably R. Baines, J. Blanchard, F. F. H. Charlton, H. Christiansen, J. Compton, Mrs. J. Cowley, D. Foreman, O. Fragman, Dr. E. Gabrielian, Prof. F. Garbari, C. Geoffroy, G. Henry, Miss M. Johnson, Dr. M. Koyuncu, F. Kummert, Dr. C. R. Lovell, E. Markus, Miss V. Matthews, Dr. T. Norman, Drs. E. & N. Özhatay, E. Pasche, Dr. P. Perry, W. Roderick, J. Rukšāns, M. Salmon, J. Sharman, Dr. A. Shmida, O. Sønderhousen, Prof. W. T. Stearn, N. Stevens, R. Suckow, N. J. Turland, Mrs. J. White, M. Young; also Dr. P. Hanelt & Dr. R. Fritsch of the Zentralinstitut für Genetik und Kulturpflanzenforschung, Gatersleben, Germany, Dr. D. Astley and his colleagues at the Institute of Horticultural Research, Wellesbourne, U.K., and the members of the 1990 Trent College Expedition to the Altai Mts. This material has admirably augmented the already very large living collection of *Allium* species cultivated at Kew, and my thanks go to the staff of the Living Collections Department whose skill in tending this valuable material facilitates the work of the botanists at the Royal Botanic Gardens. I have also benefited from the unpublished work of Henry Nielsen who prepared a survey of the economically important species of *Allium*, working on a grant from IBPGR.

For the use of the very extensive and valuable reference collection of dried material [at least 10,000 *Allium* specimens!] at Kew I thank the Keeper, Prof. G. Ll. Lucas and his staff who curate the collections, and I acknowledge with grateful thanks the assistance provided by the Directors and Keepers of the following Herbaria for allowing access to, or loans of, their herbarium specimens: AEF, ATH, B, BC, BG, BM, BRNM, CGE, E, ERE, FI, G, GB, GZU, HUJ, ISTE, JE, LD, LE, LINN, OXF, PRC, RO, S, SEV, TBI, TL, W, WAG, WSY, WU.

I also acknowledge the photographic expertise of the Media Resources Section at Kew. Examples of their fine work have been used to illustrate various species of *Allium* section *Allium*. Other photographs have been provided by the author, Dr. Phillip Cribb, Jill Cowley and Margaret Johnson.

INTRODUCTION

"The genus *Allium* can only be studied satisfactorily from living specimens; for, in Herbaria, the species of this genus, like other liliaceous plants, are seldom found to retain their characters, so as to be recognised with certainty."-George Don, 1827.

* * * * * * * *

To a large extent this statement is undoubtedly true, and much time during the two years of the IBPGR-funded taxonomic revision of section *Allium* has been spent in tracking down and collecting together living material of as many as possible of the 114 species which are recognised here. The general distribution of the section is roughly from Portugal eastwards to Central Asia and one might reasonably conclude that this region is rather well-botanised and readily accessible. However, many of the species occur in areas which are presently very difficult of access, mainly for political reasons, so that a surprisingly high number of taxa are still known from only a few collections of dried material, and in some instances only from type specimens which are apparently now non-existent. In spite of these factors, over 50% of the species have been obtained as living material [see Appendix I]; in most cases these are represented by more than one accession, and the collection which is currently housed at Kew is undoubtedly an important bank of principally wild-source material. The holdings of the genus *Allium* as a whole amount to about 1000 accessions, comprising at least 250 species out of an estimated 750.

The need for a revision of section *Allium* was already well-known, but the fact was highlighted by a study of the economically important species of *Allium* by Henry Nielsen in 1988; this project, which was based at Kew, was also financed by IBPGR. The object of the survey was to compile a list of the economically interesting species of *Allium*, together with related species, and record from the field notes on herbarium labels data such as distribution and ecology, and any further information which might be of value to researchers in other fields. However, one major drawback which Nielsen found was that 'a significant proportion of the herbarium specimens appeared to be unreliably named' and he noted that 'during this study it became clear that taxonomic revisions are badly needed for many of the species groups' (H.Nielsen, unpublished report).

The following taxonomic review of section *Allium* goes a considerable way to fulfilling the need, although inevitably there are limitations brought about by the timescale [2 years] and the already mentioned difficulties in obtaining material from certain areas. During the study it soon became apparent that it would not be possible to clarify all of the problems, and that the most useful approach would be to provide a framework in the form of a broad survey of the species of section *Allium*, drawing attention to any specific problems which could then be studied in greater detail at a more local level when time and travel permitted.

Section *Allium* contains the economically important Garlic, *A.sativum*, and *A.porrum*, the Leek, both of which are cultivated plants of some antiquity and which are not known to occur in the wild. Although 112 species are grouped together with these in the section because of certain features which they all have in common, it would be misleading to imply that they are all closely related to the Leek and Garlic. Nevertheless, some species clearly are more closely related to these crop plants and it is of value to plant breeders to be aware of such relationships. It is not possible at this stage to make a convincing infra-sectional classification since so many of the taxa are poorly known, but on page 52 I have suggested some informal groups, including one which encompasses Leek and Garlic.

The taxonomy of *Allium* undoubtedly is not an easy matter. Morphological differ-

1

entiation is rather weak and it appears that at specific level other disciplines such as palynology and cytology provide rather limited further information. However, pre-liminary biochemical studies by Harborne & Williams, using the living material available at Kew to identify the leaf flavonoids, suggest that this particular field would be worth pursuing if a wider range of material became available in future. It is admitted that the specific limits adopted in this account are fairly narrow but, until all of the taxa are known in much fuller detail in the living state, and population sur-veys have been undertaken, it is considered preferable to present this narrow view rather than to lose sight of what might be genetically interesting variants in an even larger morass of synonymity.

HISTORY OF SECTION ALLIUM

George Don, in his *Monograph of the genus Allium* (1827), divided the 139 species which he recognised into seven groups. These were taken by some later authors as representing sections, although Don did in fact formally subdivide the seven groups into Divisions and Sections and it is more likely that he was implying a subgeneric status for these major groupings. The first of Don's groups [?subgenera] was *Porrum*, an epithet used at generic level by Tournefort in 1700 and validated as a generic name under post-Linnaean nomenclature by Philip Miller in 1754; this was based on the Leek which Miller regarded as generically distinct from Onion and Garlic. Most of the subsequent treatments involving European and western Asiatic species of *Allium*, for example Regel (1875), Vvedensky (1935), Feinbrun (1943) & Omelczuk (1962) continued to recognise section *Porrum* which is comprised of those species of *Allium* possessing tricuspidate inner filaments and which appear to form a reasonably convincing assemblage. Since, however, section *Porrum* contains the type species of the genus it is correct to refer to this assemblage as section *Allium*, a position adopted by more recent authors such as Wendelbo (1971), Tscholokaschvili (1975), Wilde-Duyfjes (1976), Stearn (1978 & 1980), Pastor & Valdes (1983) & Kollmann (1984). Apart from this change of name to bring the nomenclature into line with modern practice, there have been few departures from the concept of section *Allium* [*Porrum*], although Boissier (1882) gave it only subsectional status in section *Crommyum* Webb & Berthelot. There have, however, been attempts to subdivide the section, notably by F.Hermann (1939) and Omelczuk (1962). The former recognised two subsections, *Scorodoprason*, comprising the 'flat-leaved' species, and *Oenoprason*, which housed those having subterete fistulose [hollow] leaves; this is a practical classification but almost certainly gives a false impression of species relationships. Omelczuk worked only with those species which occur in the Ukraine region and divided the section into seven series, *Vinealia*, *Scorodoprasa*, *Rotunda*, *Ampeloprasa*, *Sphaerocephala*, *Margaritacea* and *Sativa*, using features such as the bulb tunic, bulblet and bulbil characters, leaf structure and included/exserted stamens. Tzagolova (1975), working with the species from Kazachstan, also described two series, *Filidentia* and *Longicuspidata*.

It is my view that if there is to be an infra-sectional classification at all, then Omelczuk's interesting proposals would form a satisfactory basis, although in order to group all of the 114 recognised species into such a system it would be necessary to describe considerably more series than the ones currently in existence. Unfortunately at the present time many of the species are poorly known and it would be premature to propose an infra-sectional classification for section *Allium* as a whole. However, there are some fairly obvious groups within the section and these informal groups are given on page 51, following on from the list of accepted species.

ALLIUM L. section ALLIUM

Allium L., Sp. Pl. 1:295(1753)[as to spp. 1,2,10,11,12,14,15,18]. Lectotype: *A.sativum.*
Porrum Miller, Gard. Dict. Abridged 4th ed.(1754). Type: *A.porrum.*
[?subgen.] *Porrum* (Miller) G.Don, Monogr. *Allium* 4(1827).
[?sect.] *Alliotypus* Dumort., Fl. Belg. 140(1827), excl. *Cepa.* Type: *A.sativum.*
sect. *Porrum* (Miller)Reichenb. in Mössler, Gemeinn. Handb. Gewächsk. ed.2, 1:541(1827). Type: *A.porrum.*
sect. *Crommyum* subsect. *Porrum* (Miller)Webb & Berth., Phyt. Canar. 3:342(1848).
sect. *Porrum* subsect. *Scorodoprason* F.Hermann in Feddes Repert. 46:57(1939). Type: *A.scorodoprasum.*
sect. *Porrum* subsect. *Oenoprason* F.Hermann in Feddes Repert. 46:57(1939). Type: *A.vineale.*
sect. *Porrum* series *Vinealia, Scorodoprasa, Rotunda, Ampeloprasa, Sphaerocephala, Margaritacea, Sativa* Omelczuk in Ukr. Bot. Zhur. 19,3:67-68(1962). Types: *A.vineale, A.scorodoprasum, A.rotundum, A.ampeloprasum, A.sphaerocephalum, A.margaritaceum, A.sativum* respectively.

DEFINITION OF SECTION ALLIUM (MORPHOLOGY)

Outline.
Section *Allium* encompasses those species of *Allium* which have a well-developed bulb, stem (never basal) leaves, campanulate to cup-shaped (never stellate) flowers, and filaments in two distinct whorls, the outer three nearly always simple and the inner three markedly tricuspidate (rarely 5-7--cuspidate) with the anther attached to the median cusp.

Full description of section.
Bulb ovoid to subglobose; outer tunics membranous or coriaceous, sometimes splitting into parallel fibres, or reticulate-fibrous; bulblets absent to many, varying from small, subglobose, ovoid or helmet-shaped, to elongate and acuminate, brown, yellowish or deep violet, produced adjacent to the parent bulb beneath the bulb tunics, or beneath the leaf sheaths on the 'neck' of the bulb, or occasionally on stolons away from the bulb. Leaves sheathing usually the lower quarter to half of the stem but ocasionally to two thirds and in one species up to the umbel; lamina either flat and solid (usually shallowly V-shaped) or hollow (fistulose) and terete or subterete, often channelled at least in the lower part. Umbel wholly floriferous, or partly or wholly bulbilliferous. Spathe subtending the umbel usually either early caducous, with a single long-beaked valve, or 2-4-valved and persistent. Perianth campanulate to ovoid, segments often connivent. Filaments unequal, the outer three usually narrowly triangular or subulate in the upper part, simple or rarely 3-cuspidate, the inner three 3(-5-7)-cuspidate with a broad basal lamina, the median anther-bearing cusp most commonly shorter than the basal lamina and shorter than the two filiform lateral cusps, lateral cusps usually exserted from the perianth at anthesis but soon becoming contorted. Ovary with distinct nectariferous pores.

Summary of diagnostic charaters of section *Allium*

*   Bulb ovoid to globose, often with bulblets
*   Leaves sheathing stem for 1/4 to 2/3
*   Leaf lamina either solid and flat to channelled in cross-section, or hollow and semi-terete to terete
*   Spathe valve 1, long-beaked and caducous, or valves 2-4, persistent
*   Perianth campanulate to ovoid, never stellate
*   Outer 3 filaments usually simple, triangular-subulate
*   Inner 3 filaments 3(5-7)-cuspidate, the median anther-bearing cusp usually longer than the lateral sterile cusps
*   Ovary with distinct nectariferous pores

LEAF ANATOMY OF ALLIUM sect. ALLIUM

Mary Gregory, R.B.G., Kew, Richmond, Surrey, England

Most anatomical work on the genus *Allium* has been concerned with the structure of the bulb scales and their crystals; some of the major papers that include data for species in Section *Allium* are those by Horst (1909), Menz (1910), Jaccard & Frey (1928), Chartschenko (1932), Vasilevskaya (1939) and Ricci (1963); see Information from Literature for names of species.

Leaf anatomy for the genus has been studied by many workers, notably Menz (1910, 1922) and more recently Fritsch (1988, 1992a), but there is little information for species of Section *Allium*, except for those valued as foods, e.g. *A. ampeloprasum* (Levant garlic), *A. porrum* (leek), *A. sativum* (garlic), *A. scorodoprasum* (rocambole or sand leek), and *A. sphaerocephalon* (round-headed garlic or leek), and the weedy *A. vineale*.

As part of a wider study of the Alliaceae family for the series *Anatomy of the Monocotyledons* (eds. C.R.Metcalfe, D.F.Cutler and M.Gregory), and in connection with the taxonomic review of Section *Allium* by Mathew (1995), the leaf anatomy of 22 taxa of Section *Allium* is described, based mainly on material cultivated at Kew, together with information from the literature. Slides were prepared according to the methods usually employed in the Jodrell Laboratory (see Cutler 1978) and in addition some specimens were hand-sectioned and stained in toluidine blue O; they were examined under a light microscope. (See list of material examined.)

Thanks are due to Rowena Gale, Don Kirkup and Tim Lawrence for the preparation of the slides used in this work.

Leaf anatomy

The species can be divided into two distinct groups on morphological and anatomical characters. The principal anatomical characters are uniform within each group. Variations exist in the degree of thickness of epidermal walls and cuticle, extent of papilla development, number of layers of palisade tissue and number of vascular bundles, all characters that are known to vary infraspecifically with the environmental conditions.

Group I. (=Subsection Oenoprason of F.Hermann 1939).
Leaves +/- semicircular or oblong in T.S., often ribbed; vascular bundles in one ring with xylem facing centre of leaf:

Allium affine ....almost circular with 2 rounded ribs
A. curtum..almost circular with many slight ribs
A. fuscoviolaceum ....thick-crescentiform to +\- circular, ribbed
A. guttatum ssp. guttatum ....ribbed
A. guttatum ssp. sardoum ....ribbed
A. heldreichii.. ribbed
A. hierochuntinum ....ribbed
A. junceum.. ribbed
A. proponticum...thick-crescentiform; slight abaxial ribs
A. rubrovittatum.. ribbed
A. sphaerocephalon.. oblong, not ribbed
A. vineale .... ribbed

Group II. (=Subsection Scordoprason of F.Hermann 1939)
Leaves almost flat to shallowly or distinctly V-shaped in T.S., often with abaxial

keel; vascular bundles in two rows, abaxial row of large and small bundles with xylem facing adaxial surface, adaxial row of small, inversely orientated bundles:

A. ampeloprasum ....shallowly V-shaped or almost flat
A. atroviolaceum ....V-shaped
A. commutatum ....almost flat, with distinct keel
A. erubescens ....V-shaped
A. leucanthum...almost flat, without distinct keel
A. porrum ....V-shaped
A. rotundum ....V-shaped
A. sativum ....V-shaped
A. scorodoprasum ....V-shaped
A. tuncelianum ....U-shaped, without distinct keel

Group I. Leaf surface (Fig. 1a). Leaf unifacial. *Hairs*: short, simple, unicellular, thick-walled hairs present in some species, especially on ribs. *Cuticle* bearing a central longitudinal striation over most cells or occasionally (e.g. *A. sphaerocephalon*: Krahulec, 1980) a row of micropapillae. *Epidermis*: cells in regular files, longitudinally elongated, up to 5-20 or more times longer than wide, with parallel straight lateral walls and transverse to slightly oblique end walls. *Stomata* numerous, anomocytic, occurring in most files of cells, except over ribs, where these present; stomata alternating +/- regularly with epidermal cells. *Crystals* not seen.

Group I. T.S. leaf (Fig. 1b,c). *Outline* varying from oblong without ribs in *A. sphaerocephalon* to almost circular with two rounded ribs in *A. affine* and semicircular to polygonal with 7-12 or more slight ribs in other taxa examined; without distinct midrib. *Epidermal cells* +/- isodiametric or slightly wider than high or slightly higher than wide. Outer walls plus cuticle moderately to very thick, often with a central cuticular protuberance (Fig. 1c); inner walls thin to slightly thickened, radial walls thin. One or a few cells over ribs larger and thicker-walled. *Stomata*: slightly to deeply sunken, depth varying between specimens of one species (e.g. *A. vineale*). Guard cells about half to two-thirds height of epidermal cells, with small to conspicuous outer lips and small or no inner lips; lips composed of cell wall plus cuticle. *Palisade tissue* in complete ring of 1-3 layers (number varying within a species, e.g. *A. vineale*). *Spongy mesophyll* cells circular to lobed, in a few layers, increasing in size inwards but breaking down to give a large central cavity in all species.

Laticifers situated 2-3 layers below epidermis (Fig. 1c), mostly at inner boundary of palisade cells; numerous, 1 between every 1-5 palisade cells; circular to slightly polygonal in outline, thin-walled and about equal in width to palisade cells; some in contact with bundle sheath cells. Laticifers very elongated in L.S.; end walls seen only in *A. affine* and *A. guttatum* ssp. *sardoum*, where 5-sided with 5-8 large pits.

Vascular bundles in one ring in spongy mesophyll, large bundles alternating regularly or irregularly with small and medium-sized bundles, all orientated with xylem facing centre of leaf. Large bundles oval or elongated in outline and sometimes including two or three wide xylem elements in a vertical line but xylem elements more frequently not in any regular arrangement. Numbers of bundles varying both within and between species; usually one large bundle in each rib in species with ribbed outline (Fig. 1b). *Bundle sheaths* parenchymatous; sheath may be interrupted or composed of thicker-walled cells at phloem pole; some species with a cap of parenchymatous cells at xylem pole. *Crystals* not seen.

Group II. Leaf surface (Fig. 2a). Leaf pseudo-dorsiventral. *Hairs*, when present,

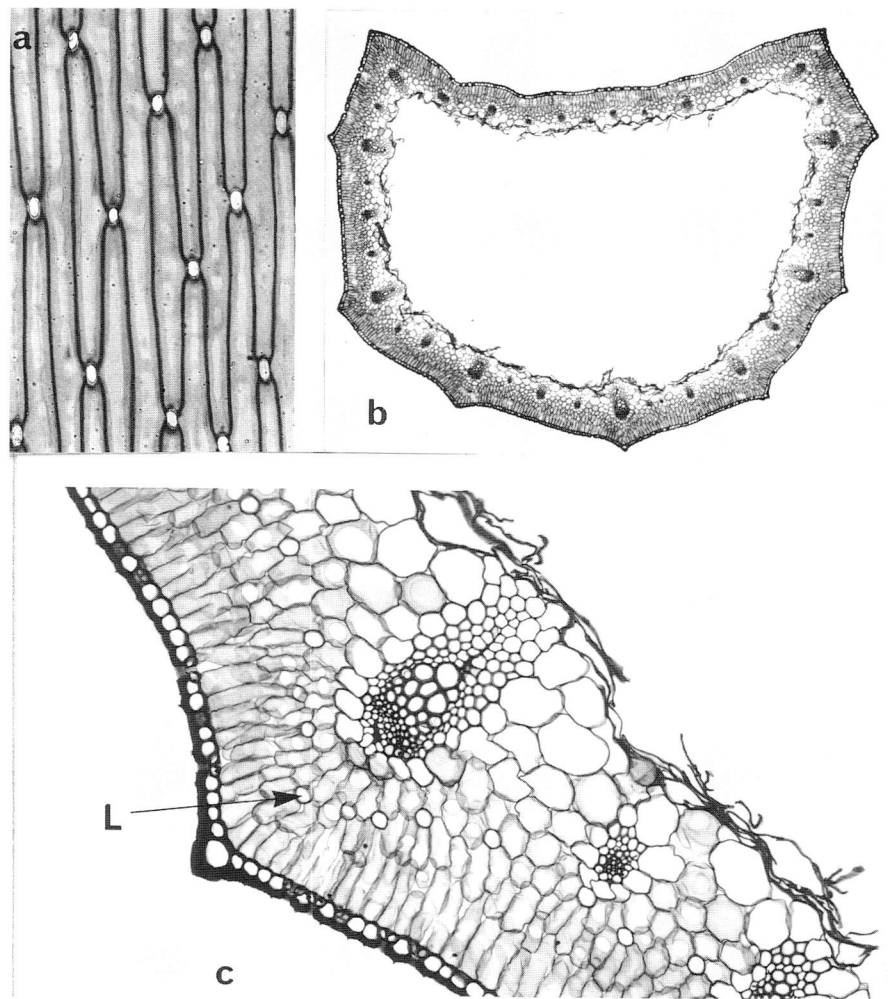

Fig. 1. Group I. a, *A. sphaerocephalon* leaf surface view x 150; b,c, *A. guttatum* ssp. *guttatum* leaf T.S., b x 25, c x 150. L=laticifer.

as in Group I. *Cuticle* bearing a central longitudinal striation over most cells or a row of micropapillae (1-2 rows in *A. commutatum*). *Epidermal cells* and stomata as in Group I; similar on both surfaces or abaxial epidermal cells slightly smaller; *stomata* +/- equally numerous on both surfaces.

Group II. T.S. leaf (Fig. 2b,c). *Outline*: almost flat with no or slight abaxial keel to V-shaped with distinct keel. *Epidermal cells* +/- isodiametric or slightly wider than high or higher than wide. Outer walls plus cuticle usually moderately thick, often domed or with a central cuticular protuberance (1-2 in *A. commutatum*); inner walls thin to slightly thickened, radial walls thin. Cells at margins and over keel taller and thicker-walled. Abaxial cells similar to adaxial or slightly smaller. *Stomata* sunken

9

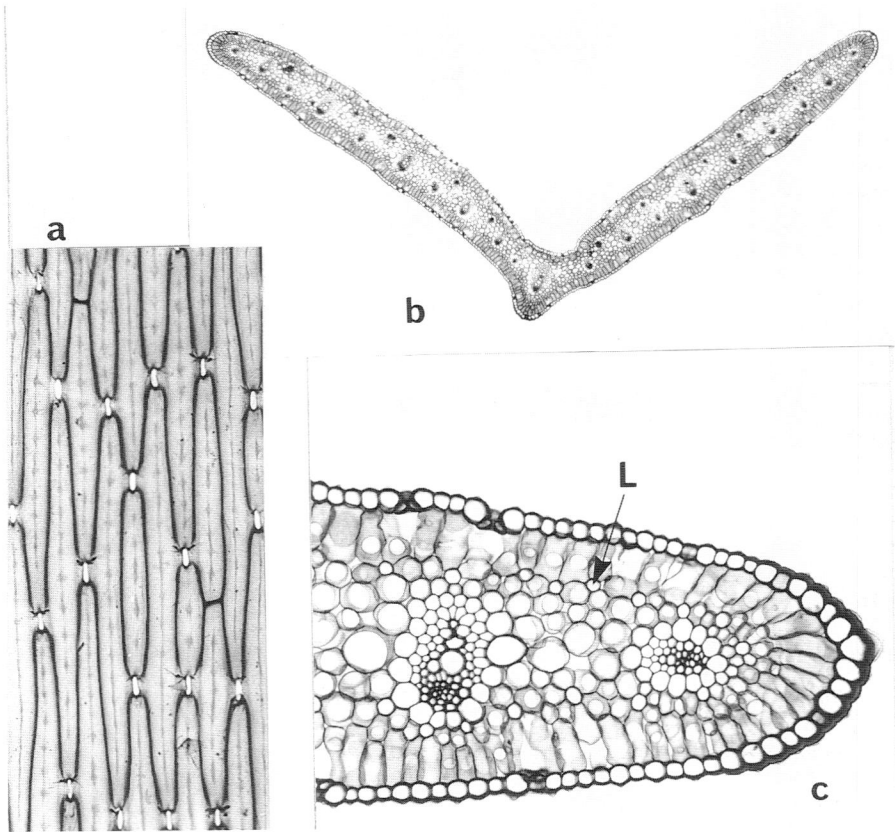

Fig. 2. Group II. a, *A. ampeloprasum* leaf surface view x 150; b,c, *A. rotundum* leaf T.S., b x 25, c, margin x 150. L=laticifer.

(superficial to slightly sunken in *A. tuncelianum*). Guard cells about half height of epidermal cells, with inconspicuous to well-developed outer lips only or with both inner and outer lips (*A. porrum, A. sativum*: Furst, 1976a). *Palisade tissue* in one layer adaxially and abaxially, except sometimes adaxially over midrib (or all cells +/- circular, as reported by Kirchner *et al*. (1934) for *A. scorodoprason*). *Spongy mesophyll* cells circular to lobed, with fairly large intercellular spaces, tending to break down between vascular bundles and forming small air canals in some species.

Laticifers situated 2-3 layers below epidermis of both surfaces (Fig. 2c), mostly at inner boundary of palisade tissue; numerous, 1 between every 1-4(-6) palisade cells; circular to slightly polygonal in outline, thin-walled and as wide as or narrower than palisade cells (but often slightly wider than palisade cells in *A. tuncelianum*). Laticifers very elongated in L.S.; end walls seen rarely, transverse to slightly oblique, with 5-6 sides and 5-8 large pits.

Vascular bundles in two rows, except sometimes only one row near margins; abaxial row consisting of large and small bundles, regularly alternating or not, all orientated with xylem facing adaxial surface (except marginal bundles may be obliquely orientated); adaxial row composed entirely of small bundles, inversely orientated (Fig. 2b). Adaxial bundles forming opposite pairs with some abaxial bundles in some species. Usually one large abaxial bundle only in midrib. Large bundles oval or elongated in outline and in most species at least some bundles with two or three wide xylem elements in a vertical line. *Bundle sheaths* parenchymatous, cells of same size as, or larger than, surrounding cells; sheath may be interrupted or composed of thicker-walled cells at phloem pole of larger vascular bundles; cap of parenchymatous cells present at xylem pole in some species. *Crystals* not seen.

Material examined [HK=Hort.Kew + Kew accession number]
Allium affine Ledeb. (HK 1976-1454); *A. ampeloprasum* L. (HK s.n.); *A. atroviolaceum* Boiss. (HK 1968-0403); *A. commutatum* Guss. (Hort. W.T.Stearn); *A. curtum* Boiss. & Gaill. (HK 1984-0154); *A. erubescens* C.Koch (Hort Mathew: Furse 7127); *A. fuscoviolaceum* Fomin (HK 1976-6059); *A. guttatum* Steven ssp. *guttatum* (HK 1976-4025); *A. guttatum* ssp. *sardoum* (Moris)Stearn (HK 1974-3476); *A. heldreichii* Boiss. (HK 1969-0804); *A. hierochuntinum* Boiss. (HK 1988-2784); *A. junceum* Smith (HK 1987-1345); *A. leucanthum* C.Koch [*A. ampeloprasum* L. var. *leucanthum* (C.Koch) Ledeb.] (HK 1969-9525); *A. proponticum* Stearn & Özhatay (HK 1976-04028); *A. rotundum* L. (HK 1976-4017); (HK 1976-1405); *A. rubrovittatum* Boiss. & Heldr. (HK 1969-8833); *A. sphaerocephalon* L. (HK 1971-3423); *A. tuncelianum* (Kollm.) N.Özhatay, B.Mathew & Şiraneci, (HK 1990-2462).

Information from literature
This is arranged by species, for the convenience of those working on disease resistance, plant breeding, ecological anatomy, etc.

A. acutiflorum Loisel.- Chartschenko 1932 (bulb scales).
A. affine Ledeb.- Tanker & Kurucu 1981 (leaf).
A. ampeloprasum L.- Drude 1933 (bulb), Horst 1909 (bulb scales), Jaccard & Frey 1928 (bulb scales), Menz 1910 (bulb, leaf), Ravololomaniraka 1972 (leaf development).
A. atroviolaceum Boiss.- Chartschenko 1932 (bulb scales).
A. borsczczowii Regel- Chartschenko 1932 (bulb scales).
A. dictyoscordum Vved.- Vasilevskaya 1939 (bulb).
A. filidens Regel- Vasilevskaya 1939 (bulb).
A. guttatum Steven (*A. margaritaceum* Sibth.)- Chartschenko 1932 (bulb scales).
A. longicuspis Regel- Vasilevskaya 1939 (bulb).
A. macrochaetum Boiss. & Hausskn.- Horst 1909 (bulb scales), Özhatay & Şiraneci 1990-2 (bulb, leaf).
A. polyanthum Schult. f.- Deloire 1980 (contractile roots), Ricci 1963 (bulb scales).
A. porrum L.- Bonnet & Boscher 1978 (leaf wax), Bugnon & Mbaya 1976 (leaf development), Cassagne & Lessire 1975 (leaf wax), Chartschenko 1932 (bulb scales), Drude 1933 (bulb), Furst 1967, 1968 (root), 1973 (leaf wax), 1976a,b (leaf), Gatin 1920 (scape), Gulyaev *et al.* 1961 (leaf), Jaccard & Frey 1928 (bulb scales), Menz 1910 (bulb, leaf, scape), Ravololomaniraka 1972 (leaf development), Ricci 1963 (bulb scales), Roth 1949 (leaf development), Thielke 1948 (leaf development), Zhestyanikova & Zykina 1976 (bulb scales).
A. pyrenaicum Costa. & Vayr.- Ricci 1963 (bulb scales).
A. rotundum L. (*A. scorodoprasum* ssp. *rotundum*) - Chartschenko 1932 (bulb scales), Jaccard & Frey 1928 (bulb scales), Kasapligil 1961 (leaf), Kirchner *et al.* 1934 (leaf), Krahulec 1977, 1980 (leaf), Menz 1910 (leaf, scape), Özhatay *et al.* 1993 (bulb, leaf). Tavel 1887 (bulb scales).

A. sativum L. (and var. *ophioscordon*)- Ahmad & Siddiqui 1987 (pharmacognosy), Bhatt 1985 (scape, root), Braecke 1921 (bulb), Chartschenko 1932 (bulb scales), Drude 1933 (bulb), Fritsch 1988 (leaf), Furst 1973 (leaf wax), 1976a,b (leaf), Gulyaev *et al.* 1961 (leaf), Huang & Sterling 1970 (bulb scales), Jaccard & Frey 1928 (bulb scales), Kothari 1979 (phloem), 1980 (leaf development), Kothari & Shah 1974a,b (development), Kothari *et al.* 1980 (laticifers), Mann 1952 (bulb, scape, leaf, root), Rahim & Fordham 1991 (leaf epidermis), Satake 1969 (pharmacognosy), Shah & Kothari 1973 (bulb), Shimoya 1970 (bulb), Thielke 1948 (leaf development), Wu & Tsai 1966 (crystals), Zhang *et al.* 1983 (laticifers), Zhestyanikova & Zykina 1976 (bulb scales).

A. scorodoprasum L. - Horst 1909 (bulb scales), Jaccard & Frey 1928 (bulb scales), Kirchner *et al.* 1934 (leaf), Krahulec 1980 (leaf), Menz 1910 (leaf), Özhatay *et al.* 1993 (bulb, leaf), Raunkiaer 1895-9 (leaf), Ricci 1963 (bulb scales).

A. sphaerocephalon L.- Chartschenko 1932 (bulb scales), Horst 1909 (bulb scales), Huang & Sterling 1970 (bulb scales), Jaccard & Frey 1928 (bulb scales), Krahulec 1980 (leaf), Kutschera & Lichtenegger 1982 (root), Ricci 1963 (bulb scales).

A. vineale L.- Boscher 1969 (bulbils), Draheim 1929 (root), Gatin 1920 (scape), Horst 1909 (bulb scales), Jaccard & Frey 1928 (bulb scales), Kirchner *et al.* 1934 (leaf), Korsmo 1954 (leaf, scape, root), Krahulec 1977, 1980 (leaf), Kutschera & Lichtenegger 1982 (root), Raunkiaer 1895-9 (leaf), Ricci 1963 (bulb scales), Stritzke & Peters 1972 (bulb).

Also Fritsch 1992b: 21 taxa of Sect. Allium (root); Fritsch 1993: 20 taxa (scape).

References

Ahmad, J. & Siddiqui, T.O. (1987). Pharmacognostical and elementological studies on *Allium sativum*, *Curcuma longa* and *Nepeta hindostana*. Hamdard Med. 30: 113-130.

Bhatt, R.P. (1985). Xylem studies in garlic (*Allium sativum* L.). Indian bot. Reporter 4: 71-72.

Bonnet, B. & Boscher, J. (1978). Importance écologique des structures superficielles de la feuille du poireau (*Allium porrum* L.) pour l'un de ses consommateurs, la teigne (*Acrolepiopsis assectella* Zell.). C.r. Acad. Sci., Paris, D, 287: 479-482.

Boscher, J. (1969) [1970]. Formation, structure et germination des bulbilles inflorescentielles de l'ail des vignes (*Allium vineale* L.). Ann. Sci. Nat., Bot., ser. 12, 10: 375-467.

Braecke, M. (1921). Étude microchimique du bulbe d'ail (*Allium sativum*). Mem. Acad. r. Belg., Cl. Sci. (Coll. 8), ser. 2, 6: 36pp. Also in: Rec. Inst. bot. Leo Errera 10: 291-318 (1922).

Bugnon, F. & Mbaya, N. (1976). La feuille "unifaciale" des Monocotyledones et son interpretation. Cas de l'*Allium porrum*. C.r. Acad. Sci., Paris, D, 282: 1507-1510.

Cassagne, C. & Lessire, R. (1975). Mouvements des cires entre la couche epicuticulaire et la cellule d'epiderme d'*Allium porrum* L. C.r. Acad. Sci., Paris, D, 280: 2537-2540.

Chartschenko, W. (1932). Verschiedene Typen des mechanischen Gewebes und der kristallinischen Ausbildungen als systematische Merkmale der Gattung *Allium*. Beih. Bot. Zbl. 50, (2): 183-206.

Cutler, D.F. (1978). Applied plant anatomy. Longman: London & New York. 103 pp.

Deloire, A. (1980). Les racines tractrices de l'*Allium polyanthum* Roem. et Schult: une étude morphologique, anatomique et histoenzymolgique. Rev. Cytol. Biol. Vég., Bot. 3: 383-390.

Draheim, W. (1929). Beiträge zur Kenntnis des Wurzelwerks von Iridaceen,

Amaryllidaceen und Liliaceen. Bot. Archiv 23: 385- 440. [Eng. summ.]

Drude, W. (1933). Beiträge zur mikroskopischen Diagnostik der Gemüse. I. Zwiebeln, Kohlarten, Tomate, Aubergine, Okra, Finocchio-Fenchel, Cardy und Artischocke. Z. Untersuch. Lebensmittel 65: 497-540.

Fritsch, R. (1988). Anatomische Untersuchungen an der Blattspreite bei *Allium* L. (*Alliaceae*). I. Arten mit einer einfachen Leitbündelreihe. Flora 181: 83-100.

Fritsch, R. (1992a). Über den Verlauf der Leitbündel in die Blattspreite bei der Gattung *Allium* L. Flora 186: 237-249.

Fritsch, R. (1992b). Zur Wurzelanatomie in der Gattung *Allium* L. (Alliaceae). Beitr. Biol. Pfl. 67: 129-160.

Fritsch, R. (1993). Anatomische Merkmale des Blütenschaftes in der Gattung *Allium* L. und ihre systematische Bedeutung. Bot. Jahrb. 115: 97-131.

Furst, G.G. (1967). Structural observations on the root system of some species of *Allium*. Byull. glavn. bot. Sada 67: 77-83. [Russ.]

Furst, G.G. (1968). Anatomical structure of the root of some species of *Allium* with varying resistance to downy mildew. Fiziologiya Immuniteta Rastenii. Akad. Nauk SSSR, Moscow. pp. 110-120. [Russ.]

Furst, G.G. (1973). Structure of waxy covering of leaves of various onion species. Byull. Glav. bot. Sada 88: 82-87. [Russ.]

Furst, G.G. (1976a). An investigation of the stomatal apparatus of the leaves of various species of *Allium* in connection with their resistance to downy mildew. Byull. Glav. bot. Sada 99: 81-91. [Russ.]

Furst, G.G. (1976b). Anatomical and histochemical properties of onion species resistant and susceptible to downy mildew. pp. 51-63 in: Fiziologiya immuniteta kul'turnikh rastenii. Akad. Nauk SSSR. Izdat. Nauka: Moskva. [Russ.]

Gatin, V.C. (1920). Recherches anatomiques sur le pedoncule et la fleur des Liliacées. Rev. gen. Bot. 32: 369-437, 460-528.

Gulyaev, V.A., Kazakova, A.A. & Syrygina, A.I. (1961). On comparative anatomy of some *Allium* L. species. Tr. Prikl. Bot. Genet. Selek. 34 (2): 14-21. [Russ.; Eng. summ.]

Hermann, F. (1939). Sectiones et subsectiones nonnullae europaeae generis *Allium*. Feddes Repert. 46: 57-58.

Horst, H. (1909). Beiträge zur vergleichenden Anatomie von Zwiebel- und Knollenschalen. Diss. Bonn. 72pp.

Huang, S.M. & Sterling, C. (1970). Laticifers in the bulb scales of *Allium*. Amer. J. Bot. 57: 1000-1003.

Jaccard, P. & Frey, A. (1928). Kristallhabitus und Ausbildungsformen des Calcium als Artmerkmal. (Ein Beitrag zur systematischen Anatomie der Gattung *Allium*.) Vierteljahrschr. naturforsch. Ges. Zürich 73, Beiblatt No. 15: 127-161. [Festschrift Hans Schinz.]

Kasapligil, B. (1961). Foliar xeromorphy of certain geophytic monocotyledons. Madrono 16: 43-70.

Korsmo, E. (1954). Anatomy of weeds. Grondahl & Sons Forlag, Oslo. 413pp.

Kothari, I.L. (1979). Morphohistogenic and anatomical studies in garlic: phloem. Proc. Indian Acad. Sci., B, 88: 219-224.

Kothari, I.L. (1980). Leaf histogenesis in garlic *Allium sativum* L. J. Indian Bot. Soc. 59: 190-195.

Kothari, I.L., Patel, J.D. & Shah, J.J. (1980). Morphohistogenic and anatomical studies in *Allium sativum* L. I. Root apical meristem. Flora 169: 230-237. II. Nuclear autolysis in phloem and laticifer. Flora 169: 238-244.

Kothari, I.L. & Shah, J.J. (1974a). Histogenesis of seed stalk and inflorescence in garlic. Phytomorphology 24: 42-48.

Kothari, I.L. & Shah, J.J. (1974b). Structure and organization of shoot apex of *Allium sativum* L. Israel J. Bot. 23: 216-222.

Krahulec, F. (1977). Poznámky k určovani československých česnecku (*Allium* L.) v nekvětoucím stavu. Zpráv. česk. bot. Spol. 12: 145-159.

Krahulec, F. (1980). Epidermal characters of *Allium* species autochthonous in Czechoslovakia: their pattern, taxonomic and ecological relationships. Preslia 52: 299-309.

Kutschera, L. & Lichtenegger, E. (1982). Wurzelatlas mitteleuropäischer Grunlandpflanzen. I. Monocotyledoneae. Gustav Fischer Verlag: Stuttgart, etc. 516 pp.

Mann, L.K. (1952). Anatomy of the garlic bulb and factors affecting bulb development. Hilgardia 21: 195-251.

Mathew, B. (1995). A taxonomic review of *Allium* Section *Allium*. Royal Botanic Gardens: Kew.

Menz, J. (1910). Beiträge zur vergleichenden Anatomie der Gattung *Allium* nebst einigen Bemerkungen über die anatomische Beziehungen zwischen Allioideae und Amaryllidoideae. Sber. Akad. Wiss. Wien, math-nat. Kl. 119 (I): 475-533.

Menz, G. (1922). Osservazioni sull'anatomia degli organi vegetativi delle specie italiane del genere *Allium* (Tourn.) L. appartenenti alla sezione *Molium* G. Don. Bull. Ist. bot. R. Univ. Sassari 1, Mem. V, 27pp.

Özhatay, N & Şiraneci, Ş. (1990-2). Comparative morphological, anatomical and preliminary chemical studies on two subspecies of *Allium macrochaetum* Boiss. et Hausskn. in Turkey. Istanbul Ecz. Fak. Mec. 26-28: 31-41.

Özhatay, N., Üstün, L. & Meriçli, A.H. (1993). Comparative morphological, karyological and chemical studies on *Allium scorodoprasum* complex in European Turkey. Istanbul Ecz. Fak. Mec. 29: 31-42.

Rahim, M.A. & Fordham, R. (1991). Effect of shade on leaf and cell size and number of epidermal cells in garlic (*Allium sativum*). Ann. Bot. 67: 167-171.

Raunkiaer, C. (1895-9). De danske blomsterplanters Naturhistorie. Vol.I. Kjøbenhavn. pp.150-6, 230-40.

Ravololomaniraka, D. (1972). Contribution à l'étude de quelques feuilles de monocotylédones Bull. Mus. natn. Hist. nat., Paris, ser. 3, No. 46, Bot. 2: 29-69.

Ricci, I. (1963). Contributo alla conoscenza tipologica dell' ossalato di calcio nel genere *Allium*. Annali bot. 27: 431-449.

Roth, I. (1949). Zur Entwicklungsgeschichte des Blattes, mit besonderer Berücksichtigung von Stipular- und Ligularbildungen. Planta 37: 299-336.

Satake, M. (1969). Microscopical anatomy of powdered garlic and detection of plant tissue fragments from the adulterated tablets of vitamin B1 derivative. Bull. natn. Inst. hygienic Sci., No. 87: 78-80. [Jap.; Eng. summ.]

Shah, J.J. & Kothari, I.L. (1973) [1974]. Histogenesis of garlic clove. Phytomorphology 23: 162-170.

Shimoya, C. (1970). Anatomia do bulbo de alho (*Allium sativum* L.) durante seu ciclo evolutivo. Rev. Ceres 17: 102-118.

Stritzke, J.F. & Peters, E.J. (1972). Anatomy of wild garlic bulbs during and subsequent to after-ripening. Weed Sci. 20: 233-237.

Tanker, N. & Kurucu, S. (1981). Leaf anatomy in relation to taxonomy in species of *Allium* found in Turkey. Q. Jl. crude Drug Res. 19: 173-179.

Tavel, F. von (1887). Die mechanischen Schutzvorrichtungen der Zwiebeln. Ber. dt. bot. Ges. 5: 438-458.

Thielke, C. (1948). Beiträge zur Entwicklungsgeschichte unifäzialer Blätter. Planta 36: 154-177.

Vasilevskaya, V.K. (1939). Systematic characters in the structure of the bulbs in species of the genus *Allium* L. pp. 174-90 in: Prezidentu Akad. Nauk SSSR Akad. V.L. Komarova k semidesyatiletyu so dnya rozhdeniya i sorokanyatiletiuy nauchnoi deyatel'nosti. [Russ.]

Wu, S.H. & Tsai, C.K. (1966). The crystals in some species of *Allium*. Acta bot. sinica 14: 115-125. [Chin.; Eng. summ.]

Zhang, W.C., Yan, W.M. & Lou, C.H. (1983). Intra- and inter- cellular changes in constitution during the development of laticiferous system in garlic scape. Acta bot. sinica 25: 8-15. [Chin.; Eng. summ.] + 4 plates.

Zhestyanikova, L.L. & Zykina, A.V. (1976). Anatomical structure of the fleshy scales of the most widely distributed species of *Allium*. Byull. Vses. Ord. Lenina Inst. Rast. N. I. Vavilova 64: 35-40. [Russ. only]

CYTOLOGY OF ALLIUM SECT. ALLIUM

Margaret A.T. Johnson. Jodrell Laboratory, Royal Botanic Gardens, Kew, Richmond, Surrey, TW9 3DS, England.

and

Neriman Özhatay. Department of Pharmaceutical Botany, Faculty of Pharmacy, Istanbul University, Beyazit 34452, Istanbul, Turkey.

The classification of section *Allium* which is adopted here recognises 115 species and several subspecies, all of which occur naturally in Europe and Asia except *A. dregeanum* which grows in South Africa. (See Map 1, p.45.)

The basic chromosome numbers for *Allium* are usually x = 8 and x = 7, the latter being restricted almost entirely to 'New World' species, with x = 8 occurring in the 'Old World' (Johnson 1982). A survey of the chromosome indexes of Fedorov (1969), Goldblatt (1981, 1984, 1985 & 1988), Moore (1973, 1974 & 1977) and Goldblatt & Johnson (1990 & 1991) together with others such as Floras (Kollmann 1984) and some papers from journals not covered by the chromosome indexes, and new data obtained at the Jodrell Laboratory (marked * in Table 1 and Appendix 1), shows that counts are known for 75 out of a total of 115 species in *Allium* section *Allium*. These counts have been extracted from 192 papers, many of which contain only one or two *Allium* section *Allium* counts.

Almost all chromosome results available for section *Allium* are based on x = 8 with the exception of one record of 2n = 14 (x = 7) for *A. heldreichii* (Alden 1976). Just 30 records are from meiotic studies, the vast majority of counts having been obtained from somatic observations, mostly since the mid 1950s, using a standard Feulgen stained squash technique (e.g. Johnson 1982). Early counts, such as those of Levan 1929, 1931, 1935 and 1940 would have been obtained from stained sections.

All the chromosome numbers recorded for each species are listed in Table 1 together with the author(s) and year of publication. A summary is given in Table 2, p. 32. Counts marked with an asterisk denote new data (not previously published) from living material at Kew (Appendix 1). Assuming that each chromosome record when given for the same species by the same author(s) in more than one paper represents a count from a different accession, there are 495 chromosome records. The 2n number ranges from 12 (one record) to a single record of 64 (8x), with both diploid (2n = 16) and tetraploid (2n = 32) counts for many species. There are 42 counts for *A. sativum* (garlic) all except one being diploid (including two aneuploid records). On the other hand, the 32 counts for *A. porrum* (leek) along with *A. kurrat* are all tetraploid (2n = 32) with the only three B chromosome records at the tetraploid level found in *A. porrum*. Within the 45 records for *A. rotundum*, a variable species, are many different chromosome numbers ranging from 2n = 16 to 2n = 64 including the only three aneuploid records at the polyploid level. From the 47 counts available for *A.sphaerocephalon*, four are tetraploid records, one is triploid, and the rest are all diploid. This species has five out of the eleven B chromosome reports at the diploid level.

Triploid records (2n = 24) are rare, just 19 having been reported; there are 22 pentaploids (2n = 40, including two aneuploids) and one heptaploid (2n = 56). These usually sterile 'odd' rather than 'even' polyploid records are to be expected in the genus, since *Allium* is well adapted to reproduction by vegetative means. There are 138 tetraploid counts which together with other polyploid records (20 3x, 22 5x, 28 6x, 1 7x and 1 8x) total 210, a surprisingly high proportion compared with the total number of counts (495). The majority of counts (285) were diploid, 2n =16 (Table 2).

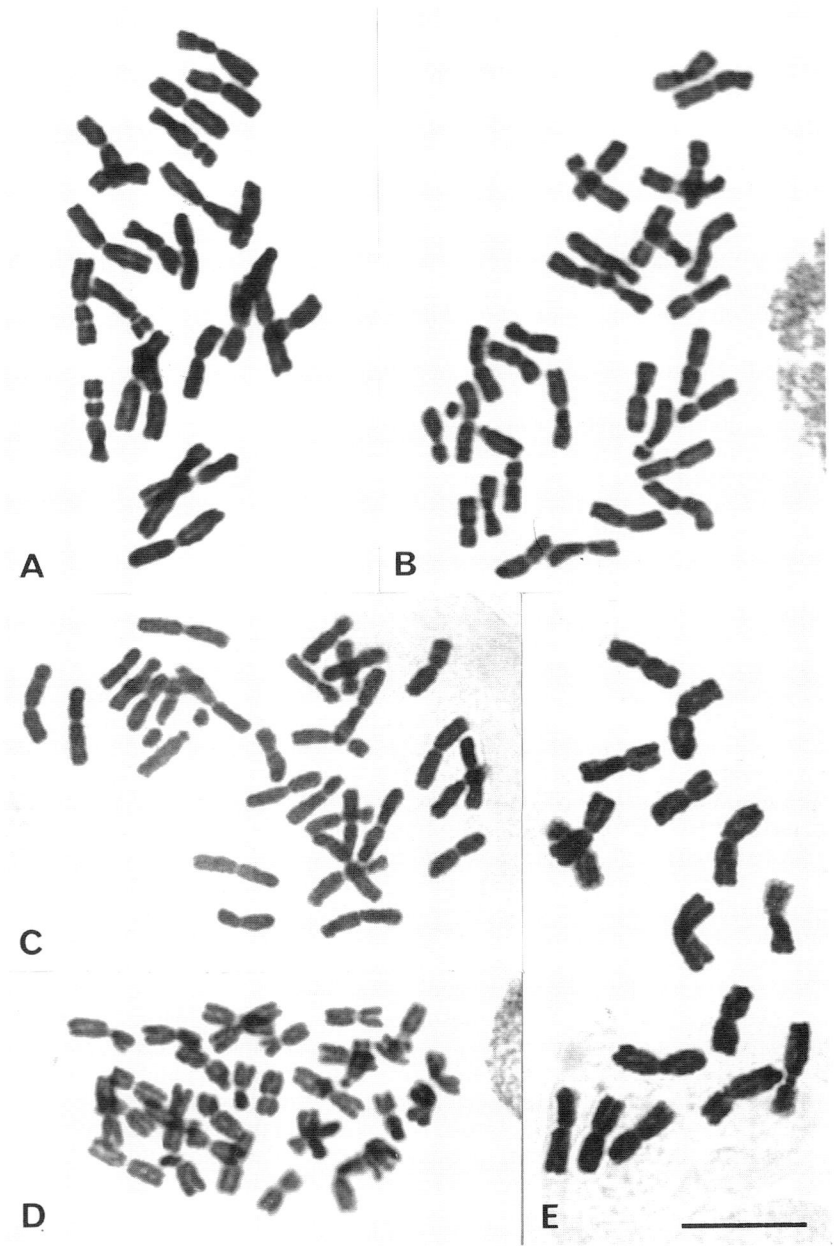

Fig. 3. Somatic cells of a *Allium atroviolaceum* (90-266) 2n=24. b *A. atroviolaceum* (90-6) 2n=32. c *A. atroviolaceum* (90-10) 2n=32. d *A. atroviolaceum* (90-2) 2n=32. e *A. amethystinum* (90-3) 2n=16. Scale bar = 10µm.

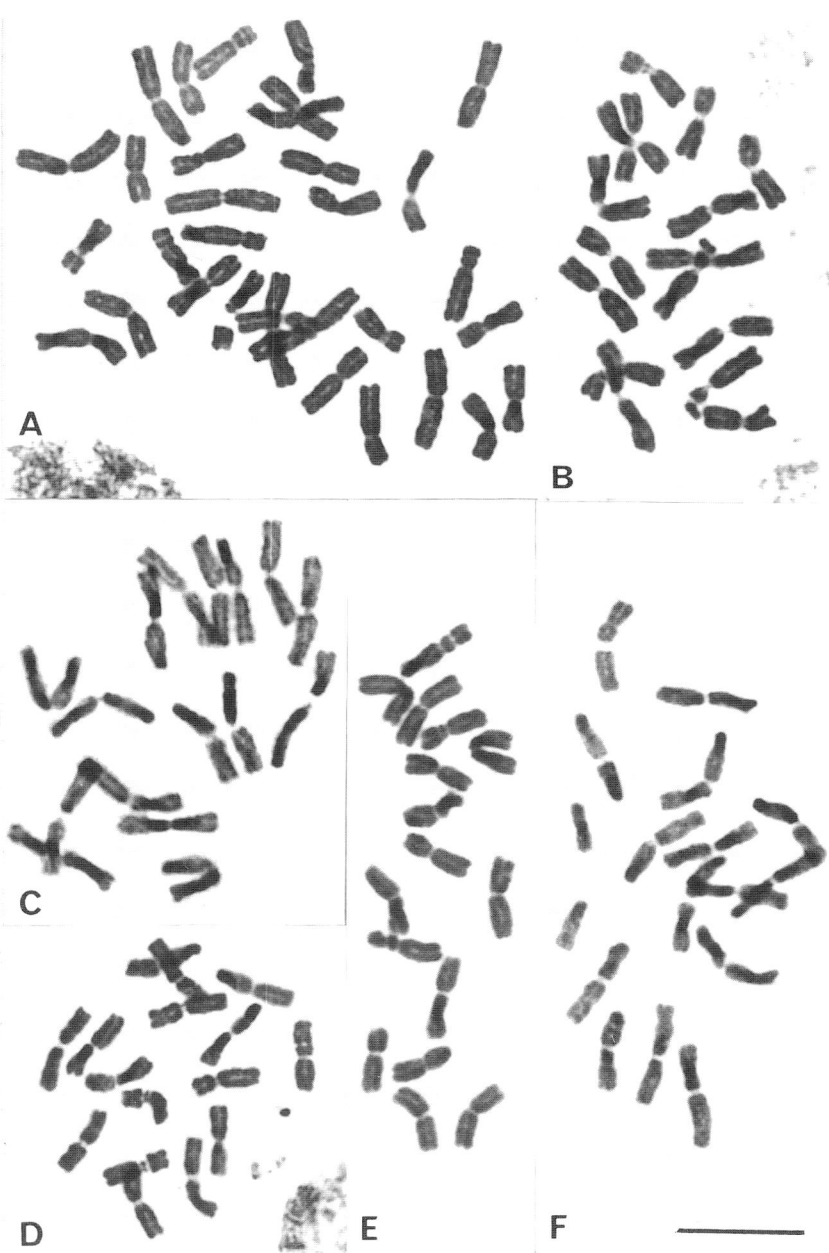

Fig. 4. Somatic cells of a *Allium ampeloprasum* (89-889) 2n=32. b *A. commutatum* (89-891) 2n=16. c *A. fuscovi-olaceum* (90-409) 2n=16. d *A. guttatum* ssp. *guttatum* (90-369) 2n=16. e *A. guttatum* ssp. *sardoum* (90-25) 2n=16. f *A. nevsehirense* (90-200) 2n=16. Scale bar = 10 μm.

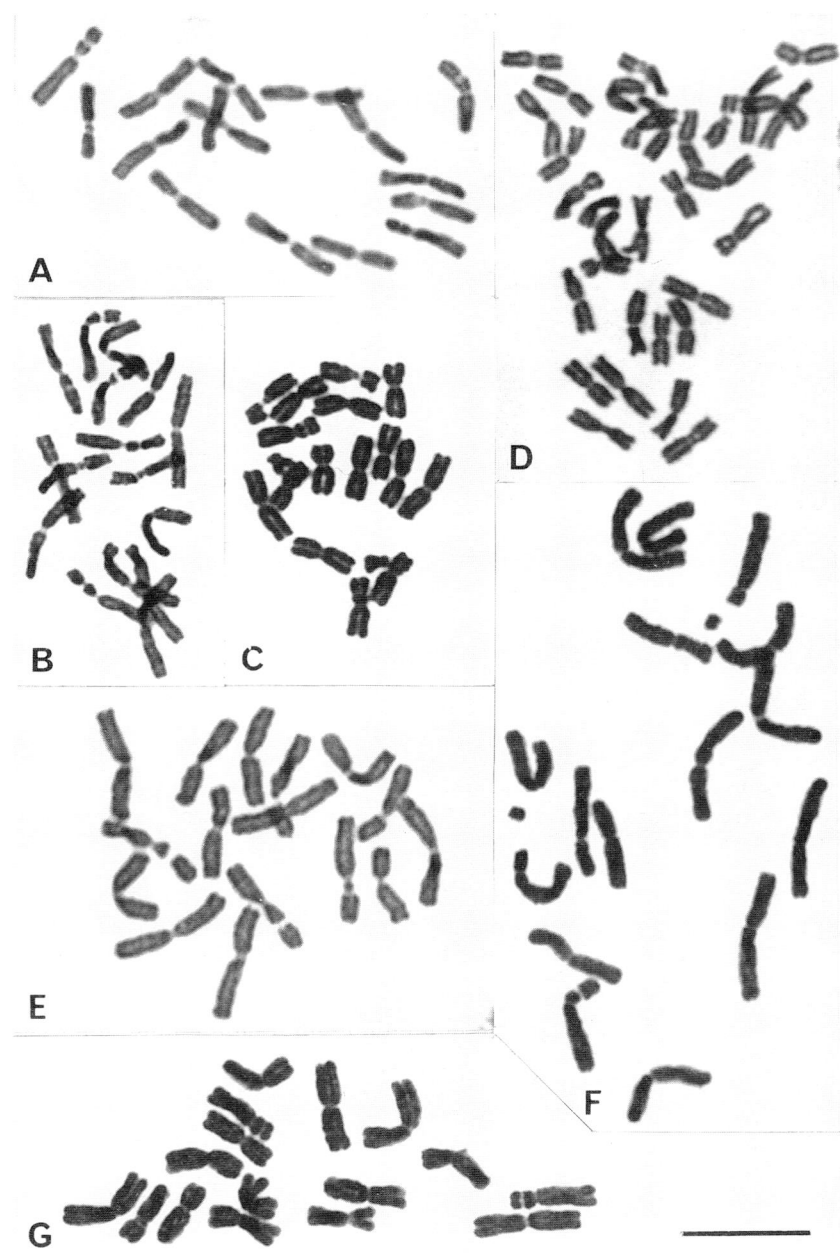

Fig. 5. Somatic cells of a *Allium rotundum* (90-52) 2n=16. b *A. rotundum* (90-184) 2n=16. c *A. rotundum* (90-194) 2n=16 + 2B. d *A. rotundum* (90-204) 2n=32. e *A. sphaerocephalon* (90-59) 2n=16. f *A. tuncelianum* (90-51) 2n=16. g *A. vineale* (89-890) 2n=16. Scale bar = 10 μm.

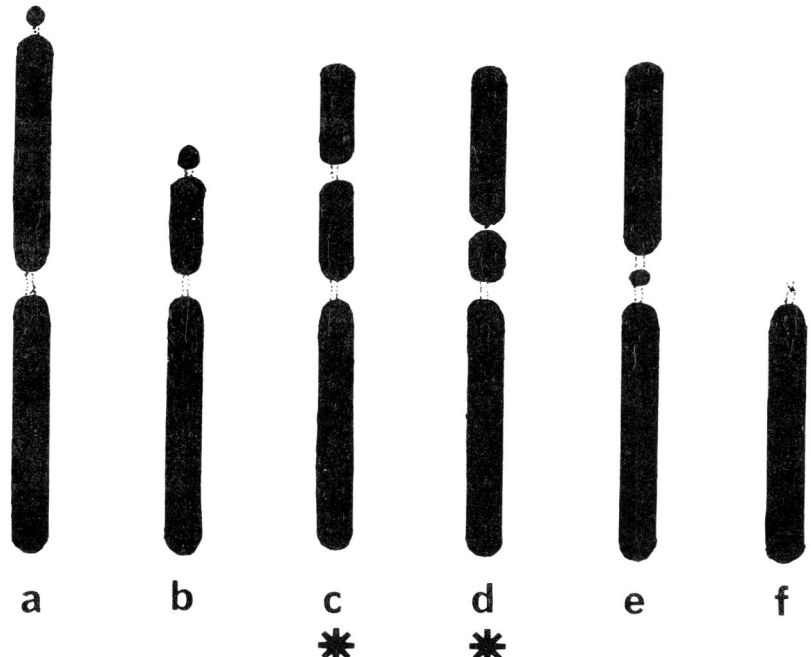

Fig. 6. Six types of nucleolar chromosomes found in *Allium* (after Ved Brat 1965). Types c and d asterisked, normally found in *Allium* section *Allium*.

Aneuploidy (±1 or 2 chromosomes) is rare, with three records by Cheshmedzhiev (1976), who found 2n = 40+1, 40+4 and 48+1 in material of *A. rotundum*, in addition to two *A. sativum* records by Banerjee (1980) and Etoh (1986), who reported 2n = 12 and 2n = 18 respectively.

Although up to 18 B chromosomes have been found in the taxa (Jones & Rees 1982), B's are not common in section *Allium*, where 7 species (16 records) with B's have been reported, the number varying from 0–4 at the diploid level, and 0–6 at the tetraploid level. Satellited chromosomes, although frequently seen in other sections of the genus, are not usually visible in section *Allium*, and no large chromosome structural changes have been reported.

The basic *Allium* section *Allium* karyotype is shown in Figs. 3–5. All species have metacentric, submetacentric or acrocentric chromosome sets. Telocentric chromosomes do not appear in this section, but are occasionally found in other sections of the genus. Although the general karyotype pattern in *Allium* is very uniform, the *Allium* section *Allium* karyotype is usually readily identified by its distinctive nucleolar chromosomes. Ved Brat (1965), recognised six types of nucleolar chromosomes in *Allium* (Fig. 6), and types c and d (asterisked) are found in *Allium* section *Allium*. These two chromosome types can usually identify a species as belonging to the section.

To aid the taxonomic revision, existing living collections of *Allium* at Kew were augmented by many new accessions particularly from Turkey which has a high number of *Allium* species. In section *Allium*, 64 taxa are represented in Turkey, 26 of which are endemic. Many of the Turkish species have been studied cytotaxonomical-

21

ly in Istanbul by Neriman Özhatay, who has published chromosome numbers together with karyotype drawings and locality data (Özhatay 1984, 1986a, 1986b, 1990 and 1993). Thirty-two new counts for section *Allium* are in press (Özhatay & Johnson).

At Kew another study utilizing the living collections has been initiated on nuclear DNA estimations. These plants are marked [DNA] on the list of living material studied. Overall chromosome size varies from species to species (see Figs. 3–5), and this is reflected in different DNA estimations, where approximately a 3-fold range was found (Chen, pers.comm.). From estimations made so far, diploids have almost exactly half the 4C DNA amount as tetraploids of the same species, and an increase in DNA is not necessarily correlated with an increase in altitude.

Although further cytological reports could reveal an increase in the degree of variation, both in ploidy level and the incidence of aneuploidy and B chromosomes, the chromosomes of *Allium* section *Allium* studied to date show a remarkably high degree of uniformity within a polyploid series based on x = 8.

Table 1. Alphabetical list of taxa in section *Allium* with chromosome numbers. In alphabetical order.
* New data (previously unpublished) obtained from living material. See Appendix 1.

Taxon ref. no.	Species	2n =	Reference Author(s)/year
11	*A. acutiflorum*	16	Garbari & Senatori 1975
82	*A. affine*	16*	Kollmann & Bothmer 1989; Özhatay & Johnson (in press); Pogosian 1983, 1988a; Tanker & Kurucu 1979
33	*A. albiflorum*		
31	*A. alibile*		
81	*A. amethystinum*	16	Cheshmedzhiev 1979; Garbari & Senatori 1975; Özhatay 1984, 1990, Tzanoudakis 1985
1	*A. ampeloprasum*	16	Arends & Laan 1979; Bothmer 1974b; Capineri 1971 [as var. *lussinense*]; Cheshmedzhiev 1971; Kollmann 1971a; Maggini & Garbari 1977; Renzoni-Cela 1964; Ved Brat 1965a, 1965b; Vosa 1977
		24*	Kollmann 1971a
		32*	Bartolo et al. 1979 [as var. *hemisphaericum*]; Bothmer 1970, 1974b, 1975a, 1975b; Cheshmedzhiev 1970, 1971; Garbari 1976; Gohil & Koul 1973, 1977, 1981; Hanelt & Ohle 1978, 1978 [as 'Perlzwiebel']; Johnson 1982; Kollmann 1971a, b [as ssp. *ampeloprasum*]; Koul & Gohil 1970a, 1973; La Cour 1945; Löve & Kjellqvist 1973; Maude 1939, 1940; Özhatay & Johnson (in press); Pastor 1982; Sopova 1972
		n=16	Gohil & Koul 1977

1	A. ampeloprasum ctd.	40	Barros Neves 1973; Bothmer 1970, 1974b, 1975a, 1975b; Kollmann 1971a, b [as ssp. ampeloprasum]; Özhatay 1984, 1990; Pastor 1982
		48	Barros Neves 1973; Bothmer 1970, 1974b, 1975b; Gohil & Koul 1973; Johnson 1982; Khoshoo et al. 1960; Kollmann 1971a; Koul & Gohil 1973; Mehra & Pandita 1978; Özhatay 1984, 1990; Pastor 1982; Stearn 1978
		n=24	Capineri et al. 1978
		56	Bothmer 1975a,b
97	A. anatolicum	16	Özhatay & Johnson (in press)
90	A. armerioides		
91	A. artemisietorum	16	Feinbrun 1950
74	A. artvinense	16	Özhatay 1986b
16	A. asirense		
42	A. asperiflorum	16	Özhatay 1986b; Özhatay & Johnson (in press)
13	A. atroviolaceum	16	Araratyan & Tonyan 1945; Gvaladze 1964; Levan 1931; Magulaev 1976; Özhatay 1984, 1986b, 1990; Pogosian 1981, 1983
		24	Cheshmedzhiev 1974; Özhatay & Johnson (in press)
		32*	Cheshmedzhiev 1970, 1974; Ghaffari 1987; Özhatay & Johnson (in press); Pogosian 1983; Sopova 1972 [as A. ampeloprasum var. atroviolaceum]; Vakhtina 1964, 1971; Vosa 1977
		40	Cheshmedzhiev 1974
		48	Cheshmedzhiev 1974
58	A. aucheri	16	Özhatay 1986b; Pogosian 1983
48	A. baeticum	32	Barros Neves 1973 [as ssp. occidentale]; Pastor 1982
93	A. barthianum		
104	A. baytopiorum	16	Özhatay & Johnson (in press)
100	A. borszczowii	16	Vakhtina 1964, 1971
3a	A. bourgeaui subsp. bourgeaui	16	Bothmer 1970, 1975a; Özhatay & Johnson (in press)
		32	Bothmer 1970, 1975a
3c	subsp. creticum	16	Bothmer 1975a
3b	subsp. cycladicum	16	Bothmer 1975a; Özhatay & Johnson (in press)

3b	A. bourgeaui subsp. cycladicum ctd.	24	Bothmer 1975a
		32	Bothmer 1975a
110	A. brevidens	16	Vakhtina 1964, 1971
		n=8	Vakhtina 1971
111	A. brevidentiforme		
45	A. calyptratum		
15	A. cappadocicum	16*	Özhatay 1986b
84	A. chamaespathum	16*	Bothmer 1974a; Tzanoudakis 1985
2	A. commutatum	16*	Bothmer 1982; Johnson 1982; Maggini & Garbari 1977
		24*	Bothmer 1984
		32	Bothmer 1975a; 1982; Garbari & Renzoni 1975
		48	Garbari & Renzoni 1975
113	A. crystallinum	16	Vakhtina 1964; 1971
		n=8	Vakhtina 1971
		32	Vakhtina 1964
67	A. curtum	16	Özhatay & Johnson (in press); Szelubsky 1950; Vosa 1977
		16 + 4B	Özhatay & Johnson (in press)
		32*	
61	A. damascenum		
96	A. deserti-syriaci		
101	A. dictyoprasum	16	Pogosian 1983, 1988b; Özhatay 1986b; Özhatay & Johnson (in press); Vosa 1977 [as A. viride]
		n=8	Kollmann 1970 [as A. viride]
		16 + 1B	Özhatay & Johnson (in press)
50	A. dictyoscordum	16	Pogosian et al. 1988
47	A. dregeanum		
37	A. drusorum		
65	A. ebusitanum	16	Cardona & Contandriopoulus 1983; Miceli & Garbari 1988a
72	A. eldivanense	16	Özhatay 1986a, 1986b
46	A. enginii	16	Özhatay & Johnson (in press)
44	A. erubescens	16*	Vakhtina 1985

44	A. erubescens ctd.	24	Pedersen & Wendelbo 1966
		32	Vakhtina 1985
27	A. esfandiarii		
79	A. ferganicum		
103	A. fethiyense	16	Özhatay & Johnson (in press)
107	A. filidens	16	Vakhtina 1964, 1971
		n=8	Vakhtina 1971
108	A. filidentiforme		
66	A. fuscoviolaceum	16	Araratyan & Tonyan 1945; Özhatay & Johnson (in press)
		16 + 1B	Özhatay 1986b
52	A. gomphrenoides	16*	Tzanoudakis 1985
83	A. gorumsense		
25	A. gramineum	16	Özhatay & Johnson (in press)
80	A. guttatum	16	Arends & Laan 1979; Buttler 1969; Chesmedzhiev 1970, 1971, 1973 [as A. margaritaceum var. margaritaceum], 1974; Mensinkai 1939, 1940 [both as A. margaritaceum var. margaritaceum]; Özhatay 1986b; Sopova 1970, 1972 [both as A. margaritaceum var. margaritaceum]; Tzanoudakis & Vosa 1988
		n=8	Cheshmedzhiev 1971, 1973 [as A. margaritaceum var. margaritaceum]; Dietrich 1972
		16 + 1B	Sopova 1970
		16 + 0 - 3B	Cheshmedzhiev 1973
		24 + 1B	Sopova 1970
		32	Mensinkai 1940; Tzanoudakis & Vosa 1988
80c	subsp. dalmaticum	32	Özhatay 1986b
80d	subsp. dilatatum		
80a	subsp. guttatum	16*	Özhatay 1986b; Tzanoudakis 1985
		32	Özhatay & Johnson (in press)
80b	subsp. sardoum	16*	Cheshmedzhiev 1970, 1974; Özhatay 1984, 1990; Tzanoudakis 1985
		n=8	Cheshmedzhiev 1974
		24	Tzanoudakis 1985
		32*	Pastor 1982
		40	Pastor 1982

25

51	A. gypsodictyum		
95	A. hamrinense		
99	A. hedgei	16	Pedersen & Wendelbo 1966
55	A. heldreichii	14	Alden 1976
		16	Johnson 1982; Levan 1935; Strid & Franzen 1981; Ved Brat 1965b
88	A. hierochuntinum	16*	
		n=8	Kollmann 1970
56	A. ilgazense	16	Özhatay 1986a, 1986b
63	A. integerrimum	16	Montmollin et al. 1986
5	A. iranicum		
57	A. jubatum	16	Özhatay 1986b
54	A. junceum	16	Miceli & Garbari 1988b
54a	subsp. junceum	16	Özhatay & Johnson (in press)
54b	subsp. tridentatum	16	Özhatay & Johnson (in press)
102	A. karyeteinii	16*	Özhatay & Johnson (in press)
76	A. kotschyi		
1b	A. kurrat	32	Kadry & Kamel 1955; Levan 1935; Vosa 1977
77	A. lehmannianum		
6	A. leucanthum	16	Levan 1931; Pogosian 1981, 1983
		32	Pogosian 1981, 1983; Vakhtina 1985
49	A. longicollum		
19	A. longicuspis	16	Dietrich 1970; Vakhtina 1964, 1971; Ved Brat 1965a, 1965b; Vosa 1977; Zakirova & Nafanailova 1988
14	A. macrochaetum	16*	Özhatay & Şiraneci 1990–2; Şiraneci 1992
75	A. makmelianum		
78	A. mareoticum		
109	A. margaritiferum		
70	A. melananthum	16	Pastor 1982
59	A. nevsehirense	16	Özhatay 1986b; Özhatay & Johnson (in press)
		32	Özhatay & Johnson (in press)
34	A. notabile		
22	A. oltense	16*	Özhatay 1986b
8	A. pardoi	48	Pastor 1982

32	A. pervestitum		
60	A. phanerantherum	n=8	Kollmann 1970 [as A. davisianum]
10	A. polyanthum	32	Löve & Kjellqvist 1973
38	A. ponticum	16	Levan 1931
		32	Özhatay 1986b
1a	A. porrum	32 *	Kadry & Kamel 1955; Kurita 1955; Levan 1931, 1935, 1940; Mehra & Pandita 1979; Murin 1964; Nakajima 1936; Nybom 1947; Pandita 1979; Pandita & Mahra 1981; Sen 1973a, 1973b; Sopova 1966; Vosa 1977
		n=16	Cheshmedzhiev 1975; Pandita & Mehra 1981
		32 + 0 - 4B	Vosa 1966
		32 + 1 - 5B	Sopova 1966
		32 + 1 - 6B	Nybom 1947
		48	Barros Neves 1973 [as A. ampeloprasum subsp. porrum]; Pandita 1979
64	A. proponticum	16	Özhatay 1984; Stearn & Özhatay 1977
64b	var. parviflorum		
64a	var. proponticum	16*	Özhatay 1990
69	A. pruinatum	16	Barros Neves 1973 [as A. ampeloprasum var. pruinatum]
4	A. pseudo-ampeloprasum	16	Özhatay 1986b
39	A. pseudo-calyptratum		
21	A. pseudo-phanerantherum		
28	A. pustulosum		
12	A. pyrenaicum	16	Ono 1935; Zhukova 1967
		32	Pastor 1982
29	A. qaradagense		
68	A. regelianum		
73	A. reuterianum	16	Özhatay & Johnson (in press)
105	A. robertianum		
23	A. rollovii	16	Özhatay 1986b
41	A. rotundum		

27

41b	subsp. *jajlae*	16	Özhatay 1986b [as *A. scorodoprasum* subsp. *jajlae*]; Özhatay & Johnson (in press); Pogosian 1983 [as *A. jajlae*]
		n=8	Dietrich 1970 [*as A. jajlae*]
		32	Özhatay & Johnson (in press)
41a	subsp. *rotundum*	16*	Cheshmedzhiev 1979 [as *A. scorodoprasum* subsp. *rotundum*]; Dietrich 1967; Löve & Löve 1982 [as *A. scorodoprasum* subsp. *rotundum*]; Majorsky et al. 1970; Özhatay 1986a, 1986b [as *A. scorodoprasum* subsp. *rotundum*]; Pogosian 1981, 1983; Sopova 1972; Weber 1929
		n=8	Dietrich 1967
		16 + 0 - 2B	Cheshmedzhiev 1976
		16 + 2B*	
		24	Cheshmedzhiev 1979 [as *A. scorodoprasum* subsp. *rotundum*]
		32*	Cheshmedzhiev 1970 [as *A. scorodoprasum* subsp. *rotundum*]; Delay 1947; Garbari & Senatori 1975 [as *A. scorodoprasum* subsp. *rotundum*]; Jacobsen & Ownbey 1977 [as *A. scorodoprasum* subsp. *rotundum*]; Johnson 1982 [as *A. scorodoprasum* subsp. *rotundum*]; Levan 1931; Özhatay 1984, 1986b [as *A. scorodoprasum* subsp. *rotundum*]; Özhatay et al. 1993; Ruiz Rejon & Sanudo 1976; Tzanoudakis 1985 [as *A. scorodoprasum* subsp. *rotundum*]
		n=16	Ruiz Rejon 1976; Cheshmedzhiev 1973
		40	Cheshmedzhiev 1970, 1971 [as subsp. *preslianum*], 1979 [as *A. scorodoprasum* subsp. *rotundum*]; Özhatay & Johnson (in press) [as *A. rotundum*]
		40 + 1	Cheshmedzhiev 1976 [as subsp. *preslianum*]
		40 + 4	Cheshmedzhiev 1976 [as *A. scorodoprasum* var. *rotundum*]
		48	Buttler 1985; Cheshmedzhiev 1971 [as var. *preslianum*]; 1979 [as *A. scorodoprasum* subsp. *rotundum*]
		48 + 1	Cheshmedzhiev 1976 [as *A. scorodoprasum* var. *rotundum*]
		64	Pastor 1982 [as *A. scorodoprasum* var. *rotundum*]

41c	subsp. *waldsteinii*	16	Vosa 1977 [as *A. waldsteinii*]
		24	Özhatay & Johnson (in press)
		32	Magulaev 1976 [as *A. waldsteinii*]; Özhatay & Johnson (in press)
53	*A. rubrovittatum*	16	Miceli & Garbari 1979; Tzanoudakis 1985
17	*A. sandrasicum*	16	Özhatay & Johnson (in press)
94	*A. sannineum*	n=8	Kollmann 1970
19a	*A. sativum*	12	Banerjee 1980
		16*	Banerjee 1980; Battaglia 1963; Cheshmedzhiev 1975; Cortes & Escalza 1986; Cortes et al. 1983; D'Amato 1950; Ferri et al. 1976; Gohil & Koul 1973b, 1981; Goto 1977; Gvaladze 1964; Katayama 1928; Khoshoo & Sharma 1957; Khoshoo et al. 1960; Koul & Gohil 1973; Koul et al. 1979; Krivenko 1936, 1938; Levan 1931, 1935; Li 1986, 1989; Maggini & Garbari 1977; Mensinkai 1939, 1940; Murin 1978; Murin & Ferakova 1973; Narain & Raina 1975; Ono 1935; Pandita 1979; Roy 1978; Sato & Kawamura 1981; Sato et al. 1980; Sen 1973a; Verma & Mittal 1978; Vosa 1977
		n=8	Etoh 1986; Koul & Gohil 1970
		n=9	Etoh 1986
		48	Khoshoo & Sharma 1957
9	*A. scaberrimum*		
89	*A. scabriflorum*	16*	Özhatay & Johnson (in press)
40	*A. scorodoprasum*	16	Cheshmedzhiev 1970, 1971 [as var. *ananthum*]; Cheshmedzhiev 1970, 1974 [as var. *scorodoprasum*]; Dietrich 1967; Halkka 1985; Hindakova & Cincura 1967; Katayama 1936; Kim 1974; Levan 1931, 1935; Löve & Löve 1982; Morinaga & Fukushima 1931; Ono 1935; Özhatay et al. 1993; Polya 1950; Semerenko 1985; Sopova 1968, 1972; Speta 1984; Tschermak-Woess 1947
		n=8	Arohonka 1982; Dietrich 1967
		24	Tschermak-Woess 1947
		32	Gadella & Kliphuis 1973; Loon 1982; Özhatay 1984 [as subsp. *scorodoprasum*].
98	*A. sinaiticum*		
42	*A. sintenisii*		
114	*A. sosnowskyanum*	16	Özhatay 1986b, 1993

62	A. sphaerocephalon	16*	Afzal-Rafii et al. 1985; Bothmer 1970; Cheshmedzhiev 1971; Dietrich 1967; Fernandes 1950; Fernandez Casas & Garcia-Villaraco 1981; Fernandez Casas et al. 1980; Guillen & Rejon 1984; Johnson 1982; Levan 1929, 1931, 1935; Loon & Oudemans 1976; Loon & Setten 1982; Loon & Snelders 1979; Magulaev 1976; Majovsky et al. 1970; Nilsson & Lassen 1971; Özhatay & Johnson (in press); Pogosian 1983; Renzoni-Cela & Garbari 1971; Scrugli & Bocchieri 1977; Sopova 1972; Strid & Franzen 1981; Tschermak-Woess 1947; Vosa 1977; Wittmann 1984
		n=8	Cheshmedzhiev 1971; Delay 1970; Dietrich 1967; Guillen & Rejon 1984; Ruiz Rejon 1976
		16 + 0 - 1B*	Pastor 1982
		16 + 1B	Guillen & Rejon 1984
		16 + 2B	Guillen & Rejon 1984; Ruiz Rejon & Sanudo 1976
		24	Loidl & Jones 1986
		32*	Martinoli 1955; Özhatay & Johnson (in press); Speta 1984
62c	subsp. *arvense*	16*	Tzanoudakis 1985
62d	subsp. *durandoi*		
62a	subsp. *sphaerocephalon*	16	Arends & Laan 1979; Barros Neves 1973 [as var. *sphaerocephalon*]; Özhatay 1984, 1990; Stearn & Özhatay 1977; Tzanoudakis 1985
		16 + 0 - 1B	Viegi & Renzoni 1981
62b	subsp. *trachypus*		
87	A. stearnianum	16*	
87a	subsp. *stearnianum*	16	Özhatay 1986b
87b	subsp. *vanense*		
71	A. stylosum	16*	Özhatay & Johnson (in press)
35	A. subnotabile		
86	A. subvineale		
7	A. talijevii		
24	A. talyschense		
26	A. trachycoleum	48	Özhatay & Johnson (in press)
18	A. truncatum	16	Kollmann 1971b [as A. ampeloprasum subsp. *truncatum*]

18	*A. truncatum* ctd.	24	Kollmann 1971b [as *A. ampeloprasum* subsp. *truncatum*]
		32	Kollmann 1971b [as *A. ampeloprasum* subsp. *truncatum*]
20	*A. tuncelianum*	16	Şiraneci 1992
106	*A. turcomanicum*	16	Vakhtina 1985
		n=16	Dietrich 1970
		32	Vakhtina 1964
115	*A. turkestanicum*	16	Vakhtina 1964
112	*A. valentinae*		
85	*A. vineale*	16*	Cheshmedzhiev 1974 [as var. *capsuliferum* and var. *compactum*]; Maggini & Garbari 1977; Özhatay 1984
		32*	Araratyan & Tonyan 1945; Barros Neves 1973 [as var. *compactum*]; Cheshmedzhiev 1970 [as var. *compactum*, var. *purshii* and var. *vineale*], 1973, 1979 [as var. *compactum*]; Dietrich 1967; Gadella & Kliphuis 1967, 1973; Hindakova 1976; Laane & Lie 1985; Levan 1931; Löve & Löve 1948; Marchi et al. 1974; Murin 1976; Natarajan 1977, 1978; Ono 1935; Özhatay 1986b; Pogosian 1983; Rohweder 1937; Scrugli 1972 [as var. *compactum*]; Skalinska et al. 1974; Sopova 1972; Speta 1984; Tischler 1934; Tzanoudakis 1985; Vosa 1977; Wittmann 1984
		n=16	Cheshmedzhiev 1973; Laane 1971
		40*	Barros Neves 1973 [as var. *compactum* and var. *nitens*]; Laane & Lie 1985; Pastor 1982
		48	Arends & Laan 1979; Pastor 1982
92	*A. vuralii*		
36	*A. wendelboi*		
30	*A. willeanum*	16	Miceli & Garbari 1988b
13a	*A. sp. cf. atroviolaceum*		

Table 2. Summary of chromosome records in *Allium* section *Allium*.

Ploidy Level	2n=	Number of Records	Number of species	Number of counts	
2x	12	1	A. sativum		
	14	1	A. heldreichii		
	16	270	70 species		
	16+0-1B	2	A. sphaerocephalon		
	16+1B	1	A. dictyoprasum		
	16+1B	1	A. fuscoviolaceum		
	16+1B	1	A. guttatum		
	16+1B	1	A. sphaerocephalon		
	16+0-2B	1	A. rotundum		
	16+2B	1	A. rotundum		
	16+2B	2	A. sphaerocephalon		Total
	16+0-3B	1	A. guttatum		285
	16+4B	1	A. curtum		diploid
	18	1	A. sativum		records
				285	
3x	24	19	19 species		
	24+1B	1	A. guttatum		
				20	
4x	32	135	22 species		
	32+0-4B	1	A. porrum		
	32+1-5B	1	A. porrum		
	32+1-6B	1	A. porrum		
				138	
5x	40	20	*5 species*		
	41	1	A. rotundum		
	44	1	A. rotundum		
				22	
6x	48	27	9 species		
	49	1	A. rotundum		
				28	
7x	56	1	A. ampeloprasum		Total
				1	210
8x	64	1	A. rotundum		polyploid
				1	records

Allium section *Allium* chromosome references.

Afzal-Rafii, Z., Vianot, J., Ramade, M. & Bourreil, M.P. (1985). Analyses des charactères caryologiques et écologiques de quelques taxons dans les massifs du Lubéron, de Lure et du Mont-Ventoux. Rev. Cytol. Biol. Vég. Bot. 8: 33–62.

Alden, B. (1976). Floristic reports from the high moutains of Pindhos, Greece. Bot. Not. 129: 297–321.

Araratayan, A.G. & Tonyan Ts. R. (1945). Chromosome numbers of some species of genus *Allium* L. Dokl. Akad. Nauk Armyanskoi SSR 2 (5): 141–433. (in Russian).

Arends J.C. & Laan, van der F.M. (1979). In: IOPB Chromosome number reports LXV. Taxon 28: 636–637.

Arohonka, T. (1982). Chromosome counts of vascular plants of the island Seili in Nauvo, SW Finland. Turun Yliopiston Biologian-Laitoksen Julkaisuja 3: 1–12.

Banerjee, N. (1980). Chromosome studies in some species of *Allium*. Proc. Indian Sci. Congr. Assoc. (IV,A) 67: 35.

Bartolo, G. Brullo, S. & Pavone, P. (1979). In: Numeri cromosomici per la flora Italiana: 617–631. Inform. Bot. Ital. 11: 149–159.

Barros Neves, J. (1973). Contribution à la connaissance cytotaxonomique des spermatophyta du Portugal. VII. *Liliaceae*. Bol. Soc. Brot. 47 (Ser. 2): 157–212.

Battaglia, E. (1963). Mutazione cromosomica e cariotipo fondamentale in *Allium sativum* L. Caryologia 16(1): 1–46.

Bothmer, R. Von. (1970). Cytological studies in *Allium*. Chromosome numbers and morphology in sect. *Allium* from Greece. Bot. Not. 123: 519–551.

_____ (1974a). Karyotype of *Allium chamaespathum* Boiss. Bot. Not. 127: 546–547.

_____ (1974b). Biosystematic studies in the *Allium ampeloprasum* complex. Op. Bot. (Lund) 34: 1–104.

_____ (1975a). Karyotype variation in *Allium bourgeaui*. Hereditas 79: 125–132.

_____ (1975b). The *Allium ampeloprasum* complex on Crete. Mitt. Bot. München 12: 267–288.

_____ (1982). Karyotype variation in *Allium commutatum* (*Liliaceae* s.l.). Plant Syst. Evol. 140: 179–189.

Buttler, K.P. (1969). Chromosomenzahlen und taxonomische Bemerkungen zu einigen Rumänischen Angiospermen. Rev. Roumaine Biol., Ser. Bot. 14: 275–282.

_____ (1985). Chromosomenzahlen von Gefänsspflanzen aus Hessen (und angrenzenden Ländern), 3. Folge. Hess. Florist. Briefe 34: 37–42.

Capineri, R. (1971). In: Numeri cromosomici per la flora Italiana. Inform. Bot. Ital. 3: 47–94.

_____, D'Amato, G. & Marchi, P. (1978). In: Numeri cromosomici per la flora Italiana: 534–583. Inform. Bot. Ital. 10: 421–465.

Cardona, M.A. & Contandriopoulos, J. (1983). In: IOPB Chromosome number reports LXXIX. Taxon 32: 323–324.

Cheshmedzhiev, I.V. (1970). A contribution to the cytosystematics of certain species of *Allium* L. from the Flora of Bulgaria. Bot. Zhurn. 55: 1100–1110.

_____ (1971). Cytosystematic study of some species from genera *Allium* L. and *Nectaroscordum* Lindl. Bot. Zhurn. 56: 1644–1657. (in Russian).

_____ (1973). To the cytotaxonomy of some Bulgarian *Allium* L. species. Bot. Zhurn. 58: 864–875.

_____ (1974). Cytotaxonomical study of certain onion species of the section *Allium*. Dokl. Bulg. Akad. Nauk. 27(8): 1109–1112

_____ (1975). A cytotaxonomic investigation of the cultivated *Allium* species in Bulgaria. Genet. Selekts. 6: 283–294.

_____ (1976). In: IOPB Chromosome number reports LIV. Taxon 25: 631–649.

_____ (1979). Karyosystematic Investigations on Species of the Genus *Allium* L. in Bulgaria. Fitologija (Sofia) 11: 40–46.

Cortes, F. & Escalza, P. (1986). Analysis of different banding patterns and late replicating regions in chromosomes of *Allium cepa*, *A. sativum* and *A. nigricans*. Genetica 71: 39–46.

_____, Gonzalez-Gil, G. & Hazen, M.J. (1983). C-Banding and sister chromatid exchanges in three species of the genus *Allium* (*A. cepa*, *A. ascalonicum* and *A. sativum*). Caryologia 36: 203–210.

D'Amato, G. (1950). Differenziazione istologica per endopoliploidia nelle radici di alcune monocotiledoni. Caryologia 3: 11–26.

Delay, C. (1947). Recherches sur la structure des noyaux quiescents chez les Phanérogames. Rev. Cytol. Cytophysiol. Vég. 9, 1–4: 169–222; 10: 1–4: 103–229.

Delay, J. (1970). Polyploidie dans les peuplements naturels. Inf. Ann. Caryosyst. Cytogenet. 4: 21–24.

Dietrich, W. (1967). Caryotypes de 46 espèces en culture (Jardin botanique de Strasbourg). Inf. Ann. Caryosyst. Cytogenet. 1: 23–26.

_____ (1970). Genre *Allium* - Espèces de collection. Inf. Ann. Caryosyst. Cytogén. 4: 29–30.

_____ (1972). Genre *Allium* - Espèces de collection. Inf. Ann. Caryosyst. Cytogén. 6: 17–20.

Etoh, T. (1986). Fertility of the garlic clones collected in Soviet central Asia. J. Jap. Soc. Hort. Sci. 55: 312–319.

Fedorov, A. (ed.) (1969). Chromosome Numbers of Flowering Plants. Leningrad, Acad. Sci. USSR.

Feinbrun, N. (1950). Chromosome counts in Palestinian *Allium* species. Palestine J. Bot. Jerusalem Ser. 5: 13–16.

Fernandes, A. (1950). Sobre a cariologia de algumas plantas da Serra do Gerês. Agron. Lusit. 12: 551–600.

Fernandes Casas, J., Sorolla, P. & Susanna, A. (1980). Numeros cromosomáticos de plantas Occidentales 64–69. Anales Jard. Bot. Madrid 36: 401–405.

_____ & Garcia-Villaraco, A. (1981). Numeros cromosomáticos de plantas Occidentales. Anales Jard. Bot. Madrid 38: 245–247.

Ferri, S., Franchi, G. & Pieri, M. (1976). Variazioni cariologiche e chimiche in *Allium sativum* L. Giorn. Bot. Ital. 110: 319–330.

Gadella, T.W.J. & Kliphuis, E. (1967). Chromosome numbers of flowering plants in the Netherlands III. Proc. Kon. Ned. Akad. Wetensch. (Ser. C) 70: 7–20.

_____ & _____ (1973). Chromosome numbers of flowering plants in the Netherlands VI. Proc. Kon. Ned. Akad. Wetensch., (Ser. C) 76: 303–311.

Garbari, F. (1976). Il genere *Allium* L. in Italia IX. *Allium sardoum* Moris. Inform. Bot. Ital. 8: 197–200.

_____ & Renzoni, G.C. (1975). Il genere *Allium* L. in Italia VII. Il caso di *A. commutatum* Guss. 32, 48. Lav. Soc. Ital. Biogeogr., n.s.5: 3–16.

_____ & Senatori, E. (1975). Il genere *Allium* L. in Italia. VI. Contributo alla citosistematica di alcune specie. Atti. Soc. Tosc. Sci. Nat. Mem. (Ser. B) 82: 1–23.

Ghaffari, S.M. (1987). Chromosome studies in some flowering plants of Iran. Rev. Cytol. Biol. Vég. Botaniste 10: 3–8.

Gohil, R.N. & Koul, A.K. (1973). Some adaptive genetic-evolutionary processes accompanying polyploidy in the Indian Alliums. Bot. Not. 126: 426–432.

_____ & _____ (1977). The cause of multivalent suppression in *Allium ampeloprasum* L. Beitr. Biol. Pflanzen 53: 473–478.

_____ & _____ (1981). In: IOPB Chromosome number reports LXXII Taxon

30: 707.

Goldblatt, P. (1981). (ed.) Index to plant chromosome numbers for 1975–1978. Missouri Botanical Garden.

_____ (1984). (ed.) Index to plant chromosome numbers for 1979–1981. Missouri Botanical Garden.

_____ (ed.) (1985). Index to plant chromosome numbers for 1982–1983. Missouri Botanical Garden.

_____ (ed.) (1988). Index to plant chromosome numbers for 1984–1985. Missouri Botanical Garden.

_____ & Johnson, D.E. (eds.) (1990). Index to plant chromosome numbers for 1986–1987. Missouri Botanical Garden.

_____ & _____ (eds.) (1991). Index to plant chromosome numbers for 1988–1989. Missouri Botanical Garden.

Moore, R.J. (ed.) (1973). Index to plant chromosome numbers. 1967–1971. Regnum Veg. 90.

_____ (ed.)(1974). Index to plant chromosome numbers for 1972. Regnum Veg. 91.

_____ (ed.)(1977). Index to plant chromosome numbers for 1973–1974. Regnum Veg. 96.

Goto, J. (1977). A technique for the microscopic observation of somatic division. Collect. & Breed. 39: 308–310.

Guillen, A. & Rejon, M.R. (1984). The B chromosome system of *Allium sphaerocephalon* L. (*Liliaceae*): types, effects and origin. Caryologia 37: 259–267.

Gvaladze, G.E. (1964). Karological studies of some species of *Allium*. Proceedings of the work of PhD students and young scientists. Institute of Botany, Academy of Sciences of Georgian SSR (Tbilisi) 2: 4 (in Russian).

Halkka, L. (1985). Chromosome counts of Finnish vascular plants. Ann. Bot. Fenn. 22: 315–317.

Hanelt, P. & Ohle, H. (1978). Die Perlzwiebeln der gaterssieben Sortiments und Bemerkungen zur Systematik und Karyologie dieser Sipp. Kulturpflanze 26: 339–348.

Hindakova, M. (1976). In: Index of chromosome numbers of Slovakian flora. Part 5. Acta. Fac. Rerum Nat. Univ. Comenianae Bot. 25: 1–18.

_____ & Cincura, F. (1967). Angaben über die Zahl und morphologie der Chromosomen einiger Pflanzenarten aus dem Territorium der Ostslowakei I. Acta Fac. Rerum Nat. Univ. Comenianae. Bot. 14: 181–227.

Jacobsen, T.D. & Ownbey, M. (1977). In: IOPB chromosome number reports LVI. Taxon 26: 257–274.

Johnson, M.A.T. (1982). Karyotypes of some Greek species of *Allium*. Ann. Mus. Goulandris 5: 107–119.

Jones, R.N. & Rees, H. (eds.) (1982). B Chromosomes. Academic Press. London.

Kadry, A. el R. & Kamel, S.A. (1955). Cytological studies in the two tetraploid species *Allium kurrat* Schweinf. and *A. porrum* L. and their hybrid. Svensk. Bot. Tidskr. 49: 314–324.

Katayama, T. (1928). The chomosome number in *Phaseolus* and *Allium*, and an observation on the size of stomata in different species of *Triticum*. Jour. Sci. Agric. Soc. ser. 3, 303: 52–54.

_____ (1936). Chromosome studies in some alliums. J. Coll. Agric. Imp. Univ. Tokyo 13: 431–441.

Khoshoo, T.N. & Sharma, V.B. (1957). Chromosome numbers of north Indian garlics. Curr. Sci. 26: 62.

_____ Atal, C.K. & Sharma, V.B. (1960). Cytotaxonomical and chemical investigations on the northwest Indian garlics. Res. Bull. Punjab Univ., n.s., Sci. 11: 37–47.

Kim, J.H. (1974). The study on scattering of chromosomes in cells - scattering of chromosomes by treatment with potassium ferricyanide solution. Korean J. Bot. 17: 113–117.

Kollmann, F. (1970). New chromosome counts in *Allium* species of Palestine and Mount Hermon. Israel J. Bot. 19: 245–248.

_____ (1971a). *Allium ampeloprasum* – A polyploid complex I. Ploidy levels. Israel J. Bot. 20: 13–20.

_____ (1971b). *Allium ampeloprasum* L. Israel. J. Bot. 20: 263–272.

_____ (1984). *Allium*, In: Davis, P.H. (ed.) Flora of Turkey 8: 98–211. Edinburgh.

_____ & Bothmer, R. Von (1989). *Allium affine* Ledeb.: taxonomy and karyotype. Israel J. Bot. 38: 47–57.

Koul, A.K. & Gohil, R.N. (1970a). Cytology of the tetraploid *Allium ampeloprasum* with Chiasma Localization. Chromosoma (Berl.) 29: 12–19.

_____ & _____ (1970b). Causes averting sexual reproduction in *Allium sativum* L. Cytologia 35: 197–202.

_____ & _____ (1973). Cytotaxonomical conspectus of the flora of Kashmir (1). Chromosome numbers of some common plants. Phyton (Horn) 15: 57–66.

_____ , _____ & Langer, A. (1979). Prospects of breeding improved garlic in the light of its genetic and breeding systems. Euphytica 28: 457–464.

Krivenko, A.A. (1936). The work of the Cytology Laboratory of Gribo Vskaya Station. In: Selection & Seed Culture of Vegetables. Sel'khozgiz. Moscow. 289–297. (in Russian).

_____ (1938). Cytological Studies of Garlic (*A. sativum*). Biol. Zh. 7: 47–68. (in Russian).

Kurita, S. (1955). Chromosome studies of several *Allium* plants. Jap. J. Genet. 30: 206–210.

Laane, M.M. (1971). Chromosome number in Norwegian vascular plant species: 6. Blyttia 29: 229–234. (in Norwegian).

_____ & Lie, T. (1985). Fremstilling av kromosompreparater med enkle metoder. Blyttia 1985: 7–15.

La Cour, L.F. (1945). In: Darlington, C.D., & Janaki Ammal, E.K. (eds.) Chromosome Atlas of Cultivated Plants. George Allen & Unwin Ltd. London.

Levan, A. (1929). Zahl und Anordnung der Chromosomen in der Meiosis in *Allium*. Hereditas 13: 80–86.

_____ (1931). Cytological Studies in *Allium*. A preliminary note. Hereditas 15: 347–356.

_____ (1935). Cytological Studies in *Allium*. VI. The chromosome morphology of some diploid species of *Allium*. Hereditas 20: 289–330.

_____ (1940). Meiosis of *Allium porrum*, a tetraploid species with chiasma localization. Hereditas 26: 454–462.

Li, Lin–chu. (1986). Chromosome observations of some plants in China. Guihaia 6: 99–105.

Li, R.-Q. (1989). Studies on karyotypes of vegetables in China. Wuhan University Press.

Loeve, A. & Kjellqvist, E. (1973). Cytotaxonomy of Spanish plants. II Monocotyledons. Lagascalia 3: 147–182.

Loidl, J. & Jones, G.H. (1986). Synaptonemal complex spreading in *Allium*. I. Triploid *A. sphaerocephalon*. Chromosoma 93: 420–428.

Loon, J.C., Van (1982). In: IOPB Chromosome number reports LXXVII. Taxon 31: 763–764.

_____ & Oudemans, J.J.M.H. (1976). Chromosome numbers of some Angiosperms of the southern USSR. Acta Bot. Neerl. 25: 329–336.

_____ & Setten, A.K. Van (1982). In: IOPB Chromosome number reports LXXVI. Taxon 31: 589–592.

_____ & Snelders, H.C.M. (1979). In: IOPB Chromosome number reports LXV. Taxon 28: 362–634.

Löve, A. & Löve, D. (1948). Chromosome numbers of Northern plant species. Repts. Dept. Agric. Univ. Inst. Appl. Sci. (Iceland) Ser. B, 3: 9–131.

_____ & _____ (1982). In: IOPB chromosome number reports LXXVI. Taxon 31: 583–587.

Maggini, F. & Garbari, F. (1977). Amounts of Ribosomal DNA in *Allium* (*Liliaceae*). Pl. Syst. Evol. 128: 201–208.

Magulaev, A.J. (1976). The chromosome numbers of flowering plants of the Northern Caucasus. (Part II). The Flora of the Northern Caucasus 2: 51–62.

Majovsky, J. (1970). Index of chromosome numbers of Slovakian flora. Part 2. Acta Fac. Rerum Nat. Univ. Comenianae Bot. 18: 45–60.

Marchi, P. Capineri, R. & D'Amato, G. (1974). In: Numeri cromosomici per la flora Italiana. Inform. Bot. Ital. 6: 303–312.

Martinoli, G. (1955). Cariologia di alcune specie del genere *Allium* (*Liliaceae*) della Sardegna. Cariologia 7: 145–156.

Maude, P.E. (1939). The Merton catalogue. A list of the chromosome numerals of species of British flowering plants. New Phytol. 38: 1–31.

_____ (1940). Chromosome numbers in some British plants. New Phytol. 39: 17–32.

Mehra, P.N. & Pandita, T.K. (1978). In: IOPB chromosome number reports LXI. Taxon 27: 375–392.

_____ & _____ (1979). In: IOPB chromosome number reports LXIV. Taxon 28: 405.

Mensinkai, S. (1939). The conception of the satellite and the nucleolus, and the behaviour of these bodies in cell division. Ann. Bot. 3: 763–793.

_____ (1940). Cytogenetic studies in the genus *Allium*. J. Genetics 39: 1–45.

Miceli, P. & Garbari, F. (1979). Cromosomi ed anatomia fogliare di quattro *Allium* diploidi di Grecia. Atti Soc. Tosc. Sci. Nat. Mem. Ser. B, 86: 37–51.

_____ & _____ (1988a). A contribution to cytotaxonomical knowledge of *Allium ebusitanum* Font Quer. Lagascalia 15 (Extra): 433–440.

_____ & _____ (1988b). Caryological aspects of *Allium willeanum* and *Allium junceum* (*Alliaceae*). Atti Soc. Tosc. Sci. Nat. Mem. Ser. B 95: 51–57.

Montmollin, B. de (1986). Étude cytotaxonomique de la flore de la Crète. III. Nombres chromosomiques. Candollea 41: 431–439.

Morinaga, T. & Fukushima, E. (1931). Chromosome numbers of cultivated plants. Bot. Mag. (Tokyo) 45: 140–145.

Murin, A. (1964). Chromosome study of *Allium porrum*. Caryologia 17: 575–578.

_____ (1976). In: Index of chromosome numbers of Slovakian flora. Part 5. Acta Fac. Rerum Nat. Univ. Comenianae Bot. 25: 1–18.

_____ (1978). In: Index of chromosome numbers of Slovakian flora. Part 6. Acta Fac. Rerum Nat. Univ. Comenianae Bot. 26: 1–42.

_____ & Ferakova, V. (1973). Karyotypes of some cultivated species of the genus *Allium*. Biologia (Bratislava) 28 (Ser. A): 65–71.

Nakajima, G. (1936). Chromosome numbers in some crops and wild angiosperms. Jap. J. Genet. 12: 211–218.

Narain, P. & Raina, S.N. (1975). Cytological assay of C–mitotic potency of colchicine obtained from *Gloriosa superba* L. Cytologia 40: 751–757.

Natarajan, G. (1977). Contribution à l'étude caryosystématique des espèces de la gualligue Languedocienne. Thesis. Academie de Montpellier, France.

_____ (1978). In: IOPB chromosome number reports LXII. Taxon 27: 519–535.

Nilsson, Q. & Lassen, P. (1971). Chromosome numbers of vascular plants from

Austria, Mallorca and Yugoslavia. Bot. Not. 124: 270–276.

Nybom, N. (1947). Accessory chromosomes in *Allium*. Hereditas 33: 571–572.

Ono, Y. (1935). Chromosome numbers in *Allium*. Jap. J. Genetics 11: 238–240.

Özhatay, N. (1984). Cytotaxonomic studies of the genus *Allium* in European Turkey and around Istanbul. III. Sect. *Allium* and Sect. *Melanocrommyum*. J. Fac. Pharm. Istanbul 20: 43–65.

―――― (1986a). Two new *Allium* species from Turkey. Notes Roy. Bot. Gard. Edinburgh 44: 147–150.

―――― (1986b). *Allium* species in North Anatolia and their chromosome numbers Doğa Türk Biyoloji Dergisi (Turkish J. Biol.) 10: 452–458.

―――― (1990). The genus *Allium* in European Turkey and around Istanbul. Ann. Mus. Goulandris 8: 115–128.

―――― (1993). *Allium* in Turkey: distribution, diversity, endemism and chromosome number. Proc. V O.P.T.I.M.A. Meeting, Istanbul. 8–15 Sept. 1986: 247–271.

―――― & Johnson, M.A.T. (in press). Some karyological remarks on Turkish *Allium* section *Allium*, *Bellevalia*, *Muscari* and *Ornithogalum* subgenus *Ornithogalum*. Bocconea 199X. Proc. VII O.P.T.I.M.A. Meeting, Bulgaria. 19–30 July 1993.

―――― & Şiraneci, Ş. (1990–92). Comparative Morphological, Anatomical and Preliminary Chemical Studies on Two subspecies of *Allium macrochaetum* Boiss. et Hausskn. in Turkey. J. Fac. Pharm. Istanbul. 26–28: 31–41.

―――― Üstün, L. & Mericli, A.H. (1993). Comparative morphological, karyological and chemical studies on *Allium scorodoprasum* complex in European Turkey. J. Fac. Pharm. Istanbul. 29: 33–42.

Pandita, T.K. (1979). Cytological investigations of some monocots of Kashmir. Ph.D. Thesis. Chandigarh.

―――― & Mehra, P.N. (1981). Cytology of Alliums of Kashmir Himalayas. III. Male meiosis. Nucleus (Calcutta) 24: 147–151.

Pastor, J. (1982). Karyology of *Allium* species from the Iberian Peninsula. Phyton, 22: 171–200.

Pedersen, K. & Wendelbo, P. (1966). Chromosome numbers of some S.W. Asian *Allium* species. Blyttia 24: 307–313.

Pogosian, A.I. (1981). Analiz morfologicheskikh parametrov khromosom nekotorykr kavkazskikh vidov roda *Allium*. L. Dokl. Akad. Nauk Armjansk. SSR (Erevan) 7: 39–58.

―――― (1983). Chromosome numbers of some species of the *Allium* (*Alliaceae*) distributed in Armenia and Iran. Bot. Zhurn. 68: 652–660. (in Russian).

―――― (1988a). Cytotaxonomical study of *Allium affine* and *Allium transcaucasicum* (*Alliaceae*). Bot. Zhurn. 73: 669–674.

―――― (1988b). Cytotaxonomical studies on *Allium dictyoprasum* C.A. Mey. and *A. viride* Grossh. (*Alliaceae*). Fl. Rastitelnost' Rastitelnye Resursy Armyansk SSR 11: 51–63.

Polya, L. (1950). Chromosome numbers of Hungarian Plants, II. (Magyarországi növényfajok kromoszómaszámai. II). Ann. Biol. Univ. Debrec. 1: 46–56.

Renzoni-Cela, G. (1964). Contributo alla cariologia della specie toscane del genere *Allium* (*Liliaceae*). Nuovo Giorn. Bot. Ital. 71: 573–574.

―――― & Garbari, F. (1971). Il genere *Allium* L. in Italia. II. Morfologia cromosomica di alcune specie. Atti Soc. Tosc. Sci. Nat. Mem. Ser. B, 78: 99–118.

Rohweder, H. (1937). Versuch zur Erfasslung des mengenmässigen Bedeckling des Darss und Zingst mit polyploiden Pflanzen. Ein Beitrag zur Bedeutung der Polyploidie bei der Eroberung neuer Lebensräume. Planta 27: 501–549.

Roy, S.C. (1978). Polymorphism in Giemsa banding patterns in *Allium sativum*. Cytologia 43: 97–100.

Ruiz Rejon, M. (1976). In: IOPB chromosome number reports LII. Taxon 25: 341–346.

―――― & Sanudo, A. (1976). Estudios cariológicos en especies españolas del orden Liliales *Allium, Lapiedra, Narcissus*. Lagascalia 6: 225–238.

Sato, S., Hizume, M. & Kawamura, S. (1980). Relationship between secondary constrictions and nucleolus organizing regions in *Allium sativum* chromosomes. Protoplasma 105: 77–85.

―――― & Kawamura, S. (1981). Cytological studies on the nucleus and the NOR carrying segments of *Allium sativum*. Cytologia 46: 781–790.

Scrugli, A. (1972). In: Numeri cromosomici per la flora Italiana. Inform. Bot. Ital. 4: 128–133.

―――― & Bocchieri, E. (1977). In: Numeri cromosomici per la flora Italiana: 348–357. Inform. Bot. Ital. 9: 127–133.

Semerenko, L.V. (1985). Chromosome numbers of some Byelorussian flora species. Bot. Zhurn. SSSR. 70: 130–132. (in Russian).

Sen, S. (1973a). Structural hybridity intra- and interspecific level in Liliales. Folia Biol. (Cracow) 21: 83–197.

―――― (1973b). Polysomaty and its significance in Liliales. Cytologia 38: 737–751.

Şiraneci, Ş. (1992). Türkiye'de yetişen *Allium macrochaetum* s.l. üzerinde sitotaksonomik araştirmalar. Firat Universitesi, XI Ulusal Biyoloji Kongresi. Elaziğ: 277–283. (in Turkish).

Skalinska, M., Malecka, J. & Izmailow, R. (1974). Further studies in chromosome numbers of Polish angiosperms. X. Acta Biol. Cracov., Ser. Bot. 17: 133–164.

Sopova, M. (1966). The nature and behaviour of supernumerary chromosomes in the *Rhizirideum* group of the genus *Allium*. Chromosoma 19: 149–158.

―――― (1968). Variation of the karyotype within the species *Allium scorodoprasum* L. Godisen Zborn. Piv.–Mat. Fak. Univ. Skopje, Biol. 20: 147–157.

―――― (1970). The occurrence and nature of B-chromosomes in a population of *Allium margaritaceum*. Godisen Zborn. Piv.–Mat. Fak. Univ. Skopje, Biol. 22: 179–187.

―――― (1972). Cytological study in the genus *Allium* from Macedonia. Godisen Zborn. Piv.–Mat. Fak. Univ. Skopje, Biol. 24: 83–102.

Speta, F. (1984). Über Oberösterreichs wildwachsende Laucharten (*Allium* L., *Alliaceae*). Linzer Biol. Beiträg 16: 45–81.

Stearn, W.T. (1978). European species of *Allium* and allied genera of *Alliaceae*: a synonymic enumeration. Ann. Mus. Goulandris 4: 83–198.

―――― & Özhatay, N. (1977). *Allium sphaerocephalon* and an allied species from European Turkey: *A. proponticum*. Ann. Mus. Goulandris 3: 45–50.

Strid, A. & Franzen, R. (1981). In: Chromosome number reports LXXII. Taxon 30: 829–842.

―――― & ―――― (1982). In: Chromosome number reports. Taxon 30: 831.

Szelubsky, R. (1950). Caryology and morphology of some Palestinian species of *Allium*. J. Bot. Jerusalem 5: 1–12.

Tanker, N. & Kurucu, S. (1979). Cytotaxonomical researches on some species of *Allium* naturally growing in Turkey. J. Fac. Pharm. Ankara 9: 1–82.

Tischler, G. (1934). Die Bedeutungen der Polyploidie für die Verbreitung der Angiospermen, erläutert an den Arten Schleswig-Holsteins, mit Ausblicken auf andere Florengebiete. Bot. Jahrb. 67: 1–36.

Tschermak-Woess, E. (1947). Über chromosome Plastizität bei Wildformen von *Allium carinatum* und anderen *Allium* - Arten aus Ostalpen. Chromosoma 3: 66–87.

Tzanoudakis, D. (1985). Chromosome studies in some species of *Allium* sect. *Allium* in Greece. Ann. Mus. Goulandris 7: 233–247.

_____ & Vosa, C.G. (1988). The cytogeographical distribution pattern of *Allium* (*Alliaceae*) in the Greek peninsula and islands. Pl. Syst. Evol. 159: 193–215.

Vakhtina, L.I. (1964). Chromosome numbers of species of onion distributed on the territory of the USSR. Bot. Zhurn. 49: 870–875. (in Russian).

_____ (1971). A comparative-karyological investigation of species of *Allium* belonging to the sections *Cepa* Prokh., *Haplostemon* Boiss. and *Allium*. Bot. Zhurn. 56: 1153–1162 (in Russian).

_____ (1985). Chromosome numbers in some species of the genus *Allium* (*Alliaceae*) in the Flora of the USSR. Bot. Zhurn. SSSR. 70(5): 700–701 (in Russian).

Ved Brat, S. (1965a). Genetic systems in *Allium* I. Chromosome variation. Chromosoma (Berl.) 16: 486–499.

_____ (1965b). Genetic systems in *Allium* III. Meiosis & breeding systems. Heredity 20: 323–339.

Verma, S.C. & Mittal, R.K. (1978). Chromosome variation in the common garlic *Allium sativum* L. Cytologia 43: 383–396.

Viegi, L. & Renzoni, G.C. (1981). In: Numeri cromosomici per la flora Italiana: 831–835. Inform. Bot. Ital. 13: 168–171.

Vosa, C.G. (1966). Seed germination and B-chromosomes in the leek (*Allium porrum*). In: Chromosomes Today. I: 24–27. Oliver & Boyd, Edinburgh & London.

_____ (1977). Heterochromatic patterns and species relationship. Nucleus 20: 33–41.

Weber, E. (1929). Entwicklungsgeschichtliche Untersuchungen über der Gattung *Allium*. Bot. Arch. 25: 1–44.

Wittmann, H. (1984). Beiträge zur Karyologie der Gattung *Allium* und zur Verbreitung der Arten im Bundesland Salzburg (Osterreich). Linzer Biol. Beitr. 16: 83–104.

Zakirova, R.O. & Nafanailova, I.I. (1988). Chromosome numbers in some species of the Kazakhstan flora. Bot. Zhurn. 73: 1493–1494.

Zhukova, P.G. (1967). Karyology of some plants, cultivated in the Arctic-Alpine Botanical Garden. In: N.A. Avrorin (ed.). Plantarum in Zonam Polarem Transportatio. II. Leningrad. (in Russian).

NOTES on FLAVONOID SURVEY

J.B.Harborne & C.A.Williams
Plant Science Laboratories, University of Reading

An exploratory survey of the flavonoid content of a selected range of species within section *Allium*, using leaf material provided by Kew, was carried out at the Plant Science Laboratories, University of Reading, by Professor J.B.Harborne and Dr Christine A.Williams. The purpose was to determine whether a study of the flavonoids could be used to improve the understanding of the classification in this large and taxonomically difficult group. For the survey, 37 taxa were chosen, representing as diverse a range of species within the section as possible but the choice was also somewhat dictated by the availability of fresh leaf material. The results were interesting and suggest that a wider survey of all the species might well shed new light on the infra-sectional relationships which are at present based solely on morphological data. Table 1 below shows the results of a survey of flavonoid aglycones, following acid hydrolysis of the leaf extracts. There is evidence, based on two dimensional chromatographic surveys of alcoholic leaf extracts, of considerable glycosidic complexity associated with these aglycones, but this was not further studied here. The three flavonols detected are kaempferol, quercetin and quercetin 3'-methyl ether (isorhamnetin). The five flavones detected are apigenin, luteolin, luteolin 3'-methyl ether (chrysoeriol), luteolin 4'-methyl ether (diosmetin) and tricetin 3',5'-dimethyl ether (tricin). In general, *Allium* species can be separated into three groups: those with only flavonols; those with flavonols and flavones; and those with only flavones.

Referring to the suggested informal groups of species (based on morphology) shown on page 51, the following points are of note:

* The "*Ampeloprasum* (Leek/Garlic) Group" species contain, on the whole, flavonols only, but *A.atroviolaceum*, which is thought to be closely allied to *A.ampeloprasum*, was found to possess flavones as well as flavonols. In the case of unexpected results such as this (and also *A.cappadocicum*) it is desirable to investigate a wider range of samples, check thoroughly the authenticity of the individuals used for the survey and then, if the findings are still at variance, reconsider the taxonomic grouping of the species concerned.

* The "*Rotundum* Group" species were found to possess both flavonols and flavones.

* The "*Sphaerocephalon* Group" of 12 species could possibly be extended on the basis of flavonoid data. Of the 5 species investigated, all contained flavones only, except for *A.curtum* which possessed the flavonol quercetin in addition to the flavones.

* The "*Guttatum* Group" species, although morphologically distinct from the "*Ampeloprasum* Group", were found to be primarily flavonol-containing only, although in the case of *A.guttatum* flavones were also present. Clearly this group requires further investigation and reconsideration if chemical data are to be incorporated into the classification.

In the case of some of the other species investigated it is impossible to comment on their placement since information is not available concerning the (supposedly) related taxa. A survey of all the species in the section is therefore necessary in order to be able to elucidate the questions raised by this preliminary study.

Table 1 Distribution of flavonoids in leaves of some species of *Allium* sect. *Allium*

Key:
Qu=Quercetin, Km=Kaempferol, Isorh=Isorhamnetin, Lu=Luteolin, Ap=Apigenin,
Chrys=Chrysoeriol, Dios=Diosmetin

Species	Flavonols			Flavones				
	Qu	Km	Isorh	Lu	Ap	Chrys	Dios	Tricin
*1. AMPELOPRASUM GROUP spp.1-25								
1. ampeloprasum 424-76.04020	+	+	-	-	-	-	-	-
2. commutatum 224-78.02387	flavonols present?			-	-	-	-	-
3. bourgeaui 211-73.01882	+	+	+	-	-	-	-	-
6. leucanthum 000-69.19525	+	+	-	-	-	-	-	-
13. atroviolaceum 424-76.04052	(+)	-	-	+	-	+	+?	-
14. macrochaetum 415-89.02744	+	+	+	-	-	-	-	-
15. cappadocicum 415-89.02746	+	+	+	+	-	+	-	-
17. sandrasicum 415-89.02741	+	+	+	-	-	-	-	-
19. longicuspis J.Ruksans s.n.	+	+	+	-	-	-	-	-
19a. sativum 'French Garlic'	+	+	-	-	-	-	-	-
20. tuncelianum 417-86.03639	+	+	-	-	-	-	-	-
25. gramineum 338-89.02183	(+)	(+)	-	-	-	-	-	-
*2. ROTUNDUM GROUP spp.(?26-)40-47								
28. pustulosum 173-85.02209	+	+	-	+	-	+	+	-
41. rotundum 424-76.04017	(+)	-	-	+	-	-	-	-
44. erubescens P.Furse 7127	+	(+)	(+)	+	-	-	+	-
47. dregeanum Karroo B.G.	+	+	+	-	-	-	-	-
*3. SPHAEROCEPHALON GROUP spp.62-73								

62. sphaerocephalon 304-75.02926	-	-	-	-	-	+	-	-
64. proponticum 424-76.04028	-	-	-	+	-	+	+?	-
66. fuscoviolaceum 179-76.06059	-	-	-	+	-	+	+?	-
67. curtum 036-84.00154	+	(+)	-	+	-	-	-	+
71. stylosum 415-89.02731	-	-	-	+	-	+	+	-

* 4.GUTTATUM GROUP
spp. 80-86

80. guttatum 552-88.04156	+	+	+	+	-	+	+	-
81. amethystinum 246-88.02011	+	+	-	-	-	-	-	-
84. chamaespathum F.Garbari s.n.	+	+	-	-	-	-	-	-
85. vineale 173-85.02176	+	+	+	-	-	-	-	-

*5. SCABRIFLORUM
GROUP
spp.88-93

88. hierochuntinum Lovell/Salmon	+	+	+	-	-	-	-	-
89. scabriflorum 496-85.05497	+	-	+	-	-	-	-	-

*6. FILIDENS GROUP
spp.106-113

107. filidens N.Stevens	-	-	-	-	-	+	-	-

*UNGROUPED SPECIES

52. gomphrenoides 361-84.03642	+	+	-	-	-	-	-	-
53. rubrovittatum 000-73.14201	-	-	-	+	-	+	+?	-
54. junceum 330-88.02745	-	-	-	-	-	-	-	+
55. heldreichii 103-69.00804	+	+	-	-	-	-	-	-
59. nevsehirense 415-89.02742	-	-	-	+	-	+	-	-
87. stearnianum 173-85.02138	+	-	-	+	-	-	-	-
101. dictyoprasum 415-89.02745	+	+	+	-	-	-	-	-
114. sosnowskyanum 293-82.02982	+	(+)	(+)	-	-	-	-	-

ECOLOGY

The majority of species grow in areas of autumn-winter-spring precipitation and summer drought, primarily in the Mediterranean and Irano-Turanian phytogeographical regions; the latter region is taken in a broad sense to extend from central Turkey south to Israel and Saudi Arabia and east to Central Asia. Few species occur in areas which experience rainfall throughout the year, for example the Euro-Siberian element in northern Turkey, and native species are almost entirely lacking from the very extensive part of continental Europe which is outside the influence of the Mediterranean.

The precise habitat details are on the whole poorly recorded and in vague terms it appears that the majority of species occur in open dryish (at flowering time) rocky places, often on limestone formations. Some species, e.g. *A.vineale*, *A.ampeloprasum* and *A.scorodoprasum*, are often associated with cultivated or abandoned fields; these are mainly species which have the ability to increase vegetatively quite rapidly by means of bulblets around the parent bulb or by bulbils produced in the umbel together with, or instead of, flowers. A few species occur in desert or semi-desert regions, for example *A.deserti-syriaci* and *A.hamrinense* from Iraq, and some are alpine or subalpine plants such as the Lebanese *A.sannineum* and Turkish *A.aucheri* & *A.reuterianum*. The altitude range for the section as a whole is from sea level to 3050 m. Records for 85 of the 114 recognised species have been obtained from herbarium material and literature and these show, albeit only approximately, that many of the species have a rather wide altitudinal range and 11 species are recorded from about sea level to 2000 m or more; not surprisingly this includes the widespread and/or 'weedy' species *A.ampeloprasum*, *A.rotundum*, *A.guttatum* and *A.vineale*. 75 species can be found in a zone between 250 m and 1550 m. 43 species occur below 500 m but only 5 of these are restricted to this low altitude belt. 26 species are recorded at or above 2000 m but only one is restricted to this zone, and even if this upper altitude zone is widened to include all those which are restricted to 1800 m or above, there are still only 4 species involved. As would be expected, the majority of the primarily low-

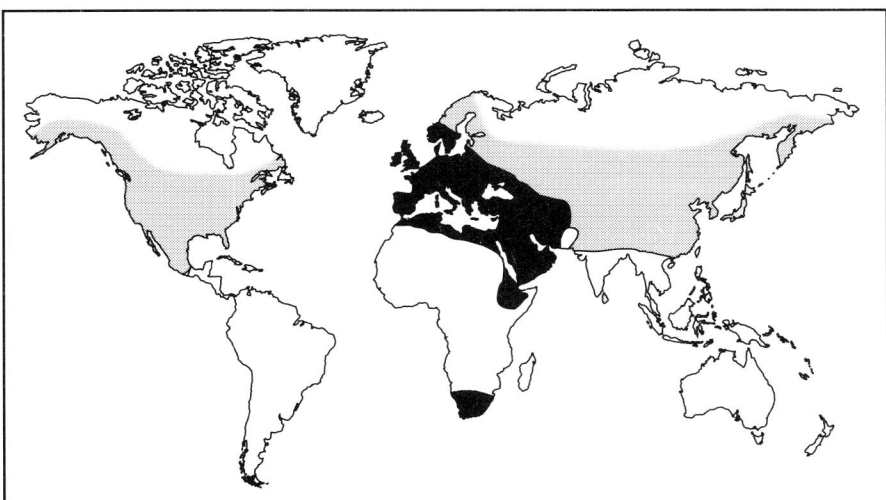

Map 1. Distribution of *Allium* species; *Allium* section *Allium* in black.

45

altitude species belong to the Mediterranean floristic element while the majority of those occurring above 1000 m might be termed Irano-Turanian species. Such comments must, however, be taken as generalisations since the records are incomplete and patchy, and the pattern is somewhat obscured by the weedy species mentioned above which are distributed over a wide altitude range.

DISTRIBUTION

The total distribution of all the species in the section extends from Portugal in the west [records of *A.ampeloprasum* in the Azores and Canary Is. may refer to naturalised plants] to Central Asia and Pakistan, approximately 75°E, and from Scandinavia in the north, southwards to Morocco, Algeria, Libya, Egypt and Saudi Arabia; there are two outlying species south of this in Ethiopia and South Africa. The extent of the range in the southeast is in Fars province of southwestern Iran.

This account recognises 114 species in section *Allium*. Approximately 30 of these are to be found in Europe, and most of these occur in the Mediterranean region; the countries which are the most species-rich are Turkey (50 spp.), former Soviet Union (37 spp.) and Iran (30 spp.), with of course some species overlapping these regions. Distributions are given as a list of countries following on from each species description.

ECONOMIC VALUE OF SPECIES IN ALLIUM SECTION ALLIUM

The Garlic, *A.sativum*, and Leek, *A.porrum*, represent the two economically most important 'species' in the section. They are undoubtedly both derived from wild ancestors, probably *A.longicuspis* and *A.ampeloprasum* respectively. Although perhaps in theory they should not qualify for specific status they do differ from these two wild species and for practical purposes it is far more convenient to refer to them as species rather than adopting a more cumbersome hierarchy which would in any case be based on a certain amount of guesswork.

GARLIC: A.sativum and A.longicuspis.

The world production of Garlic is predictably far less than that of bulb onions but nevertheless it is an important crop with a world production of over 3 million metric tons in 1989 [FAO Yearbook figures]. Its history for culinary and medicinal purposes is ancient, dating from before 3000 B.C. in Egypt, and it was also in use in China and Greece at a very early date. The uses and history are well documented, for example by Täckholm & Drar (1954) and Jones & Mann (1963), and the state of knowledge of the chemical components and their biological properties has been presented by Lutomski (1987). *A.longicuspis* is apparently not a very common plant and there are no records of it being utilised, although from personal observations of cultivated material it seems to me to have the same strong garlic smell.

LEEK: A.porrum, A.kurrat and A.ampeloprasum.

As with garlic, the leek, or something similar to it, has been cultivated since very early times but it is impossible to know whether the early records refer to *A.porrum* as we know it at the present time. Täckholm & Drar (1954) discuss this and the very similar *A.kurrat*. It is widely accepted that these two cultigens are derived from the wild *A.ampeloprasum*, although it must be said that they might equally have arisen as a result of selection from one of the closely related species (see page 70).

A.ampeloprasum itself is cultivated to a very small extent as 'Great Headed Garlic', the bulbs being utilised as an alternative to true garlic; the species is also grown as an ornamental but is of much lower aesthetic value than the popular Central Asian 'drumstick alliums'of subgenus *Melanocrommyum*.

Other useful species

A.alibile is reported to be used locally in Ethiopia for culinary purposes.

A.borszczowii. This is noted as being eaten in Baluchistan, but without precise details as to which part of the plant.

A.commutatum. The bulbs are utilised locally in the Pityusic Islands where they are reported to be ground up using pebbles and eaten with bread (Stearn, 1984).

A.macrochaetum. Said to be used as a local source of garlic in Turkey (N.Özhatay, pers. comm.).

A.rotundum. Cultivated as an ornamental to a small extent for its dense umbels of deep purple flowers.

A.scorodoprasum. In the past this has been cultivated on a small scale for flavouring purposes. The names 'Rocambole' and 'Sand Leek' are sometimes applied to this species.

A.sphaerocephalon is produced commercially in Holland for ornamental purposes and may frequently be found on sale in autumn in 'pre-packs' in garden centres and supermarkets; the interest lies in its dense umbels of dark purple flowers which are produced in summer and which can also be dried for winter decoration.

A.tuncelianum. This is reported (N.Özhatay, pers. comm.) to be used as a local source of garlic in Turkey. This is of interest since it appears to be quite closely related to *A.longicuspis* and *A.sativum* but is a fertile plant and may therefore be of value for breeding purposes.

ACCEPTED TAXA in SECT. ALLIUM

1. **A.ampeloprasum** *L.*
    1a. **A.porrum** *L.*
    1b. **A.kurrat** *Schweinf. ex Krause*
2. **A.commutatum** *Guss.*
3. **A.bourgeaui** *Rech. fil.*
    3a. subsp. **bourgeaui**
    3b. subsp. **cycladicum** *Bothmer*
    3c. subsp. **creticum** *Bothmer*
4. **A.pseudoampeloprasum** *Miscz. ex Grossh.*
5. **A.iranicum** *(Wendelbo)Wendelbo*
6. **A.leucanthum** *C.Koch*
7. **A.talijevii** *Klokov*
8. **A.pardoi** *Loscos*
9. **A.scaberrimum** *Serres*
10. **A.polyanthum** *Schultes & Schultes fil.*
11. **A.acutiflorum** *Loisel.*
12. **A.pyrenaicum** *Costa & Vayreda*
13. **A.atroviolaceum** *Boiss.*
    13a. **A.sp. Saudi Arabia cf. atroviolaceum**
14. **A.macrochaetum** *Boiss. & Hausskn.*
15. **A.cappadocicum** *Boiss.*
16. **A.asirense** *B.Mathew*
17. **A.sandrasicum** *Kollmann, N.Özhatay & Bothmer*
18. **A.truncatum** *(Feinbr.)Kollmann & D.Zohary*
19. **A.longicuspis** *Regel*
    19a. **A.sativum** *L.*
20. **A.tuncelianum** *(Kollmann)N.Ozhatay, B.Mathew & Şiraneci*
21. **A.pseudophanerantherum** *Rech.fil.*
22. **A.oltense** *Grossh.*
23. **A.rollovii** *Grossh.*
24. **A.talyschense** *Miscz. ex Grossh.*
25. **A.gramineum** *C.Koch*
26. **A.trachycoleum** *Wendelbo*
27. **A.esfandiarii** *Matine*
28. **A.pustulosum** *Boiss. & Hausskn.*
29. **A.qaradagense** *Feinbr.*
30. **A.willeanum** *Holmboe*
31. **A.alibile** *A.Rich.*
32. **A.pervestitum** *Klokov*
33. **A.albiflorum** *Omelczuk*
34. **A.notabile** *Feinbr.*
35. **A.subnotabile** *Wendelbo*
36. **A.wendelboi** *Matine*
37. **A.drusorum** *Feinbr.*
38. **A.ponticum** *Miscz. ex Grossh.*
39. **A.pseudocalyptratum** *Mouterde*
40. **A.scorodoprasum** *L.*
41. **A.rotundum** *L.*
    41a. subsp. **rotundum**

41b. subsp. **jajlae** *(Vved.)B.Mathew*
41c. subsp. **waldsteinii** *(G.Don)K.Richter*
42. **A.asperiflorum** *Miscz. ex Grossh.*
43. **A.sintenisii** *Freyn*
44. **A.erubescens** *C.Koch*
45. **A.calyptratum** *Boiss.*
46. **A.enginii** *N.Özhatay & B.Mathew*
47. **A.dregeanum** *Kunth*
48. **A.baeticum** *Boiss.*
49. **A.longicollum** *Wendelbo*
50. **A.dictyoscordum** *Vved.*
51. **A.gypsodictyum** *Vved.*
52. **A.gomphrenoides** *Boiss.*
53. **A.rubrovittatum** *Boiss. & Heldr.*
54. **A.junceum** *Smith*
54a. subsp. **junceum**
54b. subsp. **tridentatum** *Kollmann, N.Özhatay & Koyuncu*
55. **A.heldreichii** *Boiss.*
56. **A.ilgazense** *N.Özhatay*
57. **A.jubatum** *Macbride*
58. **A.aucheri** *Boiss.*
59. **A.nevsehirense** *Koyuncu & Kollmann*
60. **A.phanerantherum** *Boiss. & Hausskn.*
61. **A.damascenum** *Feinbr.*
62. **A.sphaerocephalon** *L.*
62a. subsp. **sphaerocephalon**
62b. subsp. **trachypus** *(Boiss. & Spruner)K.Richter*
62c. subsp. **arvense** *(Guss.)Arcangeli*
62d. subsp. **durandoi** *(Batt. & Trab.)Duyfjes*
63. **A.integerrimum** *Zahar.*
64. **A.proponticum** *Stearn & N.Özhatay*
64a. var. **proponticum**
64b. var. **parviflorum** *Kollmann*
65. **A.ebusitanum** *Font Quer*
66. **A.fuscoviolaceum** *Fomin*
67. **A.curtum** *Boiss. & Gaill.*
68. **A.regelianum** *A.Becker*
69. **A.pruinatum** *Link ex Sprengel*
70. **A.melananthum** *Coincy*
71. **A.stylosum** *O.Schwarz*
72. **A.eldivanense** *N.Özhatay*
73. **A.reuterianum** *Boiss.*
74. **A.artvinense** *Miscz. ex Grossh.*
75. **A.makmelianum** *Post*
76. **A.kotschyi** *Boiss.*
77. **A.lehmannianum** *Merckl.*
78. **A.mareoticum** *Bornm. & Gauba*
79. **A.ferganicum** *Vved.*
80. **A.guttatum** *Steven*
80a. subsp. **guttatum**
80b. subsp. **sardoum** *(Moris)Stearn*
80c. subsp. **dalmaticum** *(A.Kern. ex Janchen)Stearn*

80d. subsp. **dilatatum** *(Zahar.)B.Mathew*
81. **A.amethystinum** *Tausch*
82. **A.affine** *Ledeb.*
83. **A.gorumsense** *(Regel)Boiss.*
84. **A.chamaespathum** *Boiss.*
85. **A.vineale** *L.*
86. **A.subvineale** *Wendelbo*
87. **A.stearnianum** *Koyuncu, N.Özhatay & Kollmann*
    87a. subsp. **stearnianum**
    87b. subsp. **vanense** *Kollmann & Koyuncu*
88. **A.hierochuntinum** *Boiss.*
89. **A.scabriflorum** *Boiss.*
90. **A.armerioides** *Boiss.*
91. **A.artemisietorum** *Eig & Feinbr.*
92. **A.vuralii** *Kit Tan*
93. **A.barthianum** *Aschers. & Schweinf.*
94. **A.sannineum** *Gombault*
95. **A.hamrinense** *Hand.-Mazz.*
96. **A.deserti-syriaci** *Feinbr.*
97. **A.anatolicum** *N.Özhatay & B.Mathew*
98. **A.sinaiticum** *Boiss.*
99. **A.hedgei** *Wendelbo*
100. **A.borszczowii** *Regel*
101. **A.dictyoprasum** *C.A.Meyer ex Kunth*
102. **A.karyeteinii** *Post*
103. **A.fethiyense** *N.Özhatay & B.Mathew*
104. **A.baytopiorum** *Kollmann & N.Özhatay*
105. **A.robertianum** *Kollmann*
106. **A.turcomanicum** *Regel*
107. **A.filidens** *Regel*
108. **A.filidentiforme** *Vved. ex Kaschtsch.*
109. **A.margaritiferum** *Vved.*
110. **A.brevidens** *Vved.*
111. **A.brevidentiforme** *Vved.*
112. **A.valentinae** *Pavl.*
113. **A.crystallinum** *Vved.*
114. **A.sosnowskyanum** *Miscz. ex Grossh.*
115. **A.turkestanicum** *Regel*

SOME INFORMAL GROUPS OF SPECIES

Although it is impossible, with the state of knowledge of the 115 species as it is at present, to produce a meaningful infra-sectional classification, it is possible to suggest certain informal groupings of species based on character correlations.

a) "Ampeloprasum Group" [=Leek/Garlic Group]: Species 1-25. Bulb tunics membranous; bulblets yellow-brown to brown, often small, numerous and helmet-shaped; leaves solid, flat/canaliculate; spathe 1-valved with a long beak, caducous; anthers exserted.

b) "Rotundum Group": Species (?26-)40-47. Bulb tunics membranous; bulblets often dark violet; leaves solid, flat/canaliculate; spathe 1-valved, usually long-beaked, caducous; anthers usually included.

c) "Sphaerocephalon Group": Species 62-73. Bulb tunics membranous; bulblets brown; leaves hollow, terete or semi-terete; spathe with 2-4 valves, persistent; anthers exserted.

d) "Guttatum Group": Species 80-86. Bulb tunics membranous; bulblets white, grey or brown; leaves hollow, terete or semiterete; spathe 1-valved, caducous; anthers exserted.

e) "Scabriflorum Group": Species 88-93. Bulb tunics reticulate-fibrous; bulblets absent or few, brown; leaves hollow, terete or semi-terete; spathe usually splitting into 2 or more valves, persistent; anthers exserted.

f) "Filidens Group": Species 106-113. Bulb tunics reticulate-fibrous; bulblets few, brown or yellowish; leaves hollow, semi-terete; spathe 1-valved, caducous; anthers equalling perianth or exserted.

KEY TO SPECIES OF *ALLIUM* section *ALLIUM*

Notes on identification.

Several species of Allium produce wholly bulbilliferous umbels and it is therefore impossible to observe the floral characters which define sect. *Allium*. A separate key has been provided to identify such variants, which comprise group A below.

* * * * * * *

To facilitate use of the keys, attention must be given to the following details.

Bulb: the outer tunics must be intact to be sure of the true characteristics, i.e. membranous, coriaceous, reticulate-fibrous; species in which the feature is weak are keyed out twice. Note if bulblets are present, and their colour.

Leaves: note the number of leaves, their width, and the length of stem they occupy (basal third, half etc.), whether they are solid (V-shaped or flat) or fistulose (hollow, and terete or subterete); in dried material it is necessary to boil a small section of leaf to be certain of its sectional shape. Note if sheaths & lamina are scabrid or pubescent.

Spathe: note if early-caducous (as umbel expands) or persistent, the number of valves (1-4), and length of beak, if any; observe details of any bracteoles at base of pedicels.

Umbel: size, density and shape may be significant.

Flowers: note colour, shape and size of inner and outer perianth segments, and whether they are smooth or scabrid etc. on the outer surface.

Stamens: note whether anthers are exserted or included (ambiguous cases are keyed twice), and their colour (mostly yellow or pink\purple); observe features of inner filaments, especially the length of the median (anther-bearing) cusp in relation to the two lateral cusps, and the ratio of the median one to the basal undivided part of the filament; check if filaments (basal part) are pubescent\ciliate or glabrous.

NB. A few species of *Allium*, not in sect. *Allium*, have been observed to possess small lateral teeth (although not filiform cusps) on the inner filaments, thus being a possible source of confusion. *A.turkestanicum* resembles some species of sect. *Allium* in overall appearance and has been included in the section in the past; it has therefore been keyed out with them (see species no. 115); others noted are the Turkish yellow-flowered *A.microspathum* (but this differs from sect. *Allium* in its stellate perianth), and the deep reddish-purple *A.cardiostemon*, also Turkish; the latter has basal leaves, so does not resemble the species of sect. *Allium* very closely.

KEY TO GROUPS

1. Umbel consisting wholly of bulbils, flowers absent [not all species belonging to sect. *Allium*] ....Group A
   Umbels consisting of flowers only or bulbils and flowers mixed.........................2
2. Outer bulb tunics distinctly (but sometimes rather weakly) reticulate-fibrous.....3
   Outer bulb tunics membranous or coriaceous, sometimes splitting longitudinally and weakly anastomosing 4
3. Leaves non-fistulose, flattish or V-shaped (dried specimens require boiling to be certain) ....Group D
   Leaves fistulose (hollow), terete or subterete [at least in the upper part], often channelled lower down..Group F
4. Leaves non-fistulose, flattish or V-shaped (dried specimens require boiling to be certain) 5
   Leaves fistulose (hollow), terete or subterete [at least in the upper part], often channelled lower down ....Group E

5. Anthers included, or only partially exserted at anthesis...........................Group C
   Anthers fully exserted at anthesis, although sometimes only slightly so...............
   ....Group B

KEY TO GROUP A

1. Outer bulb tunics markedly reticulate-fibrous. N. America (not sect. *Allium*) ......
   ....**canadense**
   Outer bulb tunics membranous, but if splitting into fibres then not markedly retic-
   ulate. Old World [but *A.vineale* is naturalised in N America]............................2
2. Leaves basal [in flowering specimens], not sheathing the stem 3
   Leaves sheathing the lower part of the stem..4
3. Leaf solitary; stem 3-angled (not sect. *Allium*) ....**paradoxum**
   Leaves 2 or more; stem terete (not sect. *Allium*).......................................**nigrum**
4. Stem hollow and inflated, up to 4 cm diam. 5
   Stem usually solid, not inflated, rather slender...6
5. Leaves circular in section; stem widest at about the middle (not sect. *Allium*) ......
   ....**fistulosum**
   Leaves semicircular in section, stem widest below the middle (not sect. *Allium*)..
   ....**cepa**
6. Spathe persistent, with 2–4 valves 7
   Spathe falling at or before anthesis, usually 1-valved...9
7. Spathe with 2 long narrow valves, much exceeding the umbel (not sect. *Allium*) .
   ....**oleraceum**
   Spathe with 2–4 valves, equal or shorter than umbel..8
8. Leaves less than 1 mm wide and up to 18 cm long. Portugal.........**69. pruinatum**
   Leaves 1–4 mm wide and up to 30 cm long. Widespread..................................
   ....**62. sphaerocephalon**
9. Leaves hollow, terete or semi-terete, sometimes channelled in lower half.........10
   Leaves solid, flat or shallowly V-shaped...11
10. Leaves 2–3. Iran ....**86. subvineale**
    Leaves (3–)4–5(–7). Widespread in Europe/E.Med.............................**85. vineale**
11. Leaves with smooth margins...12
    Leaves with scabrid margins...14
12. Leaves 3–5(–6), 2–6(–9) mm wide. S Africa ....**47. dregeanum**
    Leaves 4–12, 5–20 mm wide 13
13. Plant of cultivation only ('Garlic'); leaves usually 6–12 ....**19a. sativum**
    Plant from natural habitats in W & C Asia; leaves 4–7 ....**19. longicuspis**
14. Plant with dark violet bulbils and bulblets; leaves 3–5 ....**40. scorodoprasum**
    Plant without, or with yellow or brown, bulblets; leaves 4–13..........................15
15. Bulb producing small subglobose or helmet-shaped bulblets beneath the tunics...
    ....**1. ampeloprasum**
    Bulb without bulblets or producing several large bulblets ('cloves') almost as
    large as the parent bulb...16
16. Plant of cultivation only ('Garlic'); leaves usually 6–12 ....**19a. sativum**
    Plant from natural habitats in W & C Asia; leaves 4–7 ....**19. longicuspis**

KEY TO GROUP B

1. Umbel consisting of bulbils and flowers...2
   Umbel consisting wholly of flowers 3
2. Leaves scabrid on margins and keel; perianth segments scabrid on keel, subacute
   to truncate...**1. ampeloprasum**

Leaves smooth or slightly scabrid; perianth segments smooth, acute ....
....**19. longicuspis**

3. Bulb tunic fibres subreticulate, anastomosing weakly and forming a long neck....
....4
Bulb tunic without obvious fibres or fibres parallel...8

4. Perianth segments pink or purple, often with a darker median vein, usually
smooth; anthers pink or violet, rarely pale yellow 5
Perianth segments white, usually scabrid or pustulose; anthers yellow, pink or
purple 6

5. Filaments of inner stamens with median cusp slightly shorter than or subequal to
the lateral cusps; leaves 1–2(–3) mm wide ....**15. cappadocicum**
Filaments of inner stamens with median cusp half as long as lateral cusps; leaves
2–5 mm wide...**14. macrochaetum**

6. Anthers pink or purple; leaves 2–5 mm wide, smooth or scabrid on the margin;
segments white with a green median vein............................**14. macrochaetum**
Anthers yellow; leaves 2–3 mm wide, scabrid at least on margins; segments white
....7

7. Perianth segments 2–3 mm long ....**28. pustulosum**
Perianth segments 3–3.5 mm long...**29. qaradagense**

8. Perianth segments pale pink to deep purple, sometimes with a darker green or
purple median vein 9
Perianth segments white or yellowish-white, sometimes with a green or purple
median vein, rarely green or greenish-white 26

9. Spathe persistent, 2-valved and visible at flowering time......**115. turkestanicum**
Spathe falling before or at anthesis, 1-valved 10

10. Perianth segments smooth 11
Perianth segments papillose or scabrid, at least on keel or margins 16

11. Plant 20–30(–45) cm; leaves 1–2(–3) mm wide 12
Plant (30–)45 cm or more; leaves 2–18 mm wide 13

12. Filaments of inner stamens with the median cusp half as long as the basal lamina
and half to two thirds as long as the lateral cusps. Saudi Arabia........................
....**16. asirense**
Filaments of inner stamens with the median cusp slightly shorter than the basal
lamina and slightly shorter than to subequal to the lateral cusps. Turkey, Iran,
Iraq, Syria...**15. cappadocicum**

13. Perianth segments deep purple or dark reddish-purple 14
Perianth segments pale pink or pinkish-mauve, sometimes with a green or purple
median vein...15

14. Median cusp of inner filaments two thirds as long as basal lamina......................
....**13. atroviolaceum**
Median cusp of inner filaments equalling basal lamina...
....**13a. sp. cf. atroviolaceum [Saudi Arabia]**

15. Leaves smooth; perianth segments 2.5–3 mm long; median cusp of inner filaments
equal to or longer than the basal lamina....................................**20. tuncelianum**
Leaves with scabrid margins; perianth segments 5 mm long; median cusp of inner
filaments half as long as the basal lamina ....**21. pseudophanerantherum**

16. Leaves smooth...17
Leaves minutely scabrid, at least on the margins..20

17. Flowers deep purple 18
Flowers pale pink or pale pinkish-purple...19

18. Leaves 8–15 mm wide; perianth segments 2.5–4 mm long; inner segments trun-

cate or emarginate..**18. truncatum**

Leaves 2–7 mm wide; perianth segments 4–5 mm long; inner segments obtuse ...
....**38. ponticum**

19. Leaves (4–)6–8 mm wide; filaments ciliate at base; anthers violet; median cusp
of inner filaments half to two thirds as long as lateral cusps...........................
....**4. pseudoampeloprasum**

Leaves 8–10 mm wide; filaments glabrous; anthers yellow; median cusp of inner
filaments two thirds to three quarters as long as lateral cusps.........**5. iranicum**

20. Leaves 3 mm wide; umbel c.3 cm diam. Iran.................................**27. esfandiarii**

Leaves usually 5–25 mm wide; umbel 3–8 cm diam..21

21. Anthers partially or slightly exserted; bulblets dark purple.............**41. rotundum**

Anthers fully exserted; bulblets brownish or yellow 22

22. Outer filaments tricuspidate; outer segments densely and minutely papillose all
over outer surface, truncate, emarginate or obtuse ....**2. commutatum**

Outer filaments simple; outer segments sparsely papillose or scabrid, especially
on the keel, acute, subacute or obtuse..23

23. Plant 30–65 cm; leaves 2–4 in number, 2–6 mm wide; umbel 1.5–3 cm diam.;
spathe with beak to 2 cm long; anthers only slightly exserted. Saudi Arabia,
Ethiopia ....**31. alibile**

Plant 45–200 cm; leaves 4–13 in number, 5–25(–50) mm wide; umbel 3–8 cm
diam.; spathe with beak to 15 cm long; anthers well exserted 24

24. Perianth segments dark purple; tunics becoming fibrous with age ....
....**13. atroviolaceum**

Perianth segments pink or pale red; tunics membranous 25

25. Bulblets few, large [up to 1.5 cm long], compressed- ovoid; outer perianth segments papillose or scabrid-ciliate; median cusp of inner filaments a quarter to a
third as long as lateral cusps..**3. bourgeaui**

Bulblets many, small, helmet-shaped or subglobose; outer perianth segments with
large sparse papillae on keel; median cusp of inner filaments half as long as lateral cusps...**1. ampeloprasum**

26. Outer perianth segments smooth, even on the keel, but sometimes with scabrid
margins..27

Outer perianth segments scabrid or papillose, at least on the keel....................34

27. Leaves 2, glabrous, c.2 mm wide; plant c.25 cm in height. Syria ...**37. drusorum**

Leaves 2–15, glabrous or scabrid, 2–25(–50) mm wide; plant (20–)35–200 cm in
height; if leaves 2, and 2 mm wide, then scabrid, and plant at least 50 cm tall ..
....28

28. Leaves glabrous; umbel 5–8 cm diam.; anthers pinkish– purple; median cusp of
inner filaments equal to or longer than the basal lamina..........**20. tuncelianum**

Leaves usually scabrid or papillose, at least on the margins [if glabrous then perianth only 2–2.5 mm long]; umbel 1.5–6 cm diam.; anthers yellow or brown;
median cusp of inner filaments half to two thirds as long as lateral cusps......29

29. Leaves densely pilose-scabrid, at least on margins, usually 2–3 in number, and
1–3 mm wide 30

Leaves scabrid but not densely so, up to 7 in number, 2.5–12 mm wide [if only
2–3 leaves, either at least 7 mm wide or only very slightly scabrid]...............31

30. Perianth segments 2–3 mm long, smooth or scabrid only on the keel; leaves 2–3
mm wide, scabrid on margins, keel and veins ....**22. oltense**

Perianth segments 3–3.5 mm long, smooth with scabrid margins; leaves 1–2 mm
wide, scabrid-dentate on margins only ....**23. rollovii**

31. Perianth yellowish-white; plant 45–50 cm in height. Ukraine.............**7. talijevii**

Perianth segments white with a green median vein; plant 50–120 cm 32

32. Outer perianth segments 2–2.5 mm long.....................................**17. sandrasicum**
    Outer perianth segments 3–4.2 mm long 33
33. Outer perianth segments 4–4.2 mm long, inner 2.8–3.5 mm; leaves 7–10 mm
    wide. Spain ....**8. pardoi**
    Outer and inner perianth segments subequal, 3–3.5 mm long; leaves 3–9 mm
    wide. Transcaucasus, Iran, Saudi Arabia....................................**6. leucanthum**
34. Leaves densely pilose-scabrid on the margins and veins, usually 2–5 in number,
    2–5 mm wide 35
    Leaves glabrous or scabrid, but not densely so, 2–15 in number, 2–50 mm wide..
   39
35. Perianth segments 2–3.5 mm long 36
    Perianth segments 4–6 mm long 38
36. Outer perianth segments 2–2.5 mm long, smooth or very slightly pustulose on the
    keel, obtuse, green with white margins. NE Turkey ....**22. oltense**
    Outer perianth segments 2–3.5 mm long, pustulose to scabrid- pustulose, acute,
    apiculate, white...37
37. Perianth segments 2–3 mm long. C & C-S Turkey ....**28. pustulosum**
    Perianth segments 3–3.5 mm long. Iraq ....**29. qaradagense**
38. Bulblets dark purple; median cusp of inner filaments usually a quarter to a third as
    long as the lateral cusps...**39. pseudocalyptratum**
    Bulblets brown or yellowish; median cusp of inner filaments half as long as to
    subequal to the lateral cusps..**26. trachycoleum**
39. Leaves glabrous, (4–)6–8 mm wide; perianth segments acute ....
    ....**4. pseudoampeloprasum**
    Leaves usually scabrid or denticulate but if smooth then 8–15 mm wide and peri-
    anth segments obtuse 40
40. Leaves up to 5 mm wide at widest point...41
    Leaves more than 5 mm wide at the widest point..48
41. Pedicels very densely and conspicuously papillose-scabrid ....**34. notabile**
    Pedicels smooth or slightly scabrid at apex 42
42. Bulblets dark purple; median cusp of inner filaments usually a quarter to a third as
    long as lateral cusps 43
    Bulblets yellowish or brownish; median cusp of inner filaments usually half as
    long as to subequal to the lateral cusps (sometimes a third as long in 38.**alibile**)
   44
43. Perianth segments 4.5–5 mm long, white with a green median vein. Lebanon,
    Israel ....**39. pseudocalyptratum**
    Perianth segments 3–4 mm long, yellowish. Ukraine ....**32. pervestitum**
44. Median cusp of inner filaments a quarter to a third as long as the basal lamina;
    anthers only partially exserted ....**25. gramineum**
    Median cusp of inner filaments half as long as to subequal to the basal lamina;
    anthers fully exserted..45
45. Perianth segments 3.5–6 mm long; median cusp of inner filaments a third to half
    as long as basal lamina 46
    Perianth segments 2–3.5 mm long; median cusp of inner filaments two thirds as
    long as to slightly shorter than the basal lamina..47
46. Plant (12–)30–65 cm; anthers very slightly exserted; perianth segments 3.5–4 mm
    long; median cusp of inner filaments a third as long as the basal lamina. Saudi
    Arabia, Ethiopia ....**31. alibile**
    Plant (50–)60–100 cm; perianth segments 4–6 mm; median cusp of inner fila-
    ments half as long as the basal lamina. Israel, N & E to Turkey & Iran ....
    ....**26. trachycoleum**

47. Outer perianth segments 2–3 mm long, obtuse, truncate or emarginate, papillose on outside. Cyprus ....**30. willeanum**
    Outer perianth segments 3–3.5 mm long, acute or subacute, smooth or very slightly scabrid on keel. Transcaucasus, Iran ....**6. leucanthum**
48. Bulblets dark purple; leaves 2.5–6 mm wide...............................**32. pervestitum**
    Bulblets, if produced, yellowish or pale brown; leaves up to 50 mm wide 49
49. Outer filaments tricuspidate; outer segments densely minutely papillose all over the outer surface, truncate, emarginate or obtuse ....**2. commutatum**
    Outer filaments simple; outer segments smooth, or scabrid, especially on the keel, acute, subacute or obtuse 50
50. Filaments glabrous; leaves sheathing three quarters of the stem; flowers yellowish-white; plant 45–50 cm. Ukraine...**7. talijevii**
    Filaments ciliate at base; leaves sheathing a quarter to a half of the stem; flowers white, often with a green or purplish median vein on the segments, or green; plant (40–)45–200 cm. 51
51. Perianth segments green; bulblets borne on distinct stolons away from the bulb ..
    ....**18. truncatum**
    Perianth segments white, often with green or purplish median vein; bulblets carried adjacent to the bulb beneath the tunics and leaf sheaths 52
52. Inner filaments with the median cusp two thirds as long as to subequal to the basal lamina; perianth segments smooth or very slightly scabrid on the keel, or furnished with sparse low papillae 53
    Inner filaments with the median cusp a quarter to half as long as the basal lamina; perianth segments conspicuously scabrid or papillose 54
53. Perianth segments smooth or very slightly scabrid on keel, 3–3.5 mm long; median cusp of inner filaments two thirds to three quarters as long as lateral cusps ..
    ....**6. leucanthum**
    Perianth segments with sparse low papillae, 3–4.5 mm long; median cusp of inner filaments half as long as lateral cusps. SE France ....**9. scaberrimum**
54. Bulblets few, large [up to 1.5 cm long], compressed-ovoid; outer perianth segments papillose or scabrid-ciliate; median cusp of inner filaments a quarter to a third as long as lateral cusps; flowers pale green ....**3. bourgeaui**
    Bulblets many, small, helmet-shaped or subglobose; outer perianth segments with large sparse papillae on keel; median cusp of inner filaments half as long as lateral cusps; flowers white with a green or purple median vein...........................
    ....**1. ampeloprasum**

KEY TO GROUP C

1. Perianth segments pink to deep purple, usually with a darker median vein 2
    Perianth segments white or greenish, sometimes with a pink, purple or green median vein, rarely suffused very pale pink or purple in the later stages........15
2. Umbel consisting of bulbils and flowers..3
    Umbel consisting wholly of flowers 4
3. Flowers purple; leaves scabrid on the margins. Europe ....**40. scorodoprasum**
    Flowers pink/white; leaves glabrous. S Africa ....**47. dregeanum**
4. Outer perianth segments furnished with very long papillae, or densely shaggy-scabrid or very coarsely papillose..5
    Outer perianth segments rather finely scabrid or warted, or finely papillose or smooth..6
5. Outer perianth segments with long (to 1 mm) papillae on the keel, 7–10 mm long. E -C Turkey ....**43. sintenisii**

Outer perianth segments shaggy-scabrid or coarsely papillose, 5–6 mm long. C-N & NE Turkey ....**42. asperiflorum**

6. Umbel conical; stem 10–20 cm, flexuose; inner perianth segments truncate. C-S Turkey ....**46. enginii**

Umbel spherical or hemispherical; stem 15–90 cm, straight; inner perianth segments usually acute, obtuse or rounded, sometimes retuse or emarginate, but if truncate-rounded then flowers pale pink 7

7. Outer perianth segments usually acuminate, but if acute then markedly outward-curved at tips; inner segments acuminate or acute 8

Outer and inner perianth segments not acuminate and not markedly outward-curved at the tips 9

8. Flowers 6–9 mm long; outer segments acuminate, erect; inner segments acuminate. S.France, Corsica, NW Italy ....**11. acutiflorum**

Flowers 5–7(–9) mm long; outer segments acuminate or acute, curving outwards at apex; inner segments acute. Caucasus, Crimea, Iran ....**44. erubescens**

9. Perianth segments +/- smooth and sometimes also shiny, 3–4 mm long; anthers usually violet...10

Perianth segments scabrid or papillose, at least on the keel, 3–8 mm long; anthers usually yellow 11

10. Median cusp of inner filaments half as long as the basal lamina. Saudi Arabia..... ....**16. asirense**

Median cusp of inner filaments about equal to basal lamina. Turkey, Iran, Iraq, Syria ....**15. cappadocicum**

11. Perianth segments 3–4 mm long; umbel rather lax; pedicels very slender. Caucasus, NE Turkey ....**38. ponticum**

Perianth segments usually 4–8 mm long; umbel usually very dense.................12

12. Inner perianth segments usually distinctly longer than outer; bulb producing many dark, often blackish-violet bulblets beneath the tunics; flowers pink to dark purple ....**41. rotundum**

Inner perianth segments subequal to or shorter than the outer; bulb producing brown or yellowish bulblets; flowers pale pink...13

13. Leaves with glabrous margins; perianth segments 5–8 mm long; median cusp of inner filaments half as long as basal lamina and two thirds as long as lateral cusps. S Africa...**47. dregeanum**

Leaves with scabrid or denticulate margins; perianth segments 3.5–5 mm long; median cusp of inner filaments a third as long as the basal lamina and a third to two thirds as long as the lateral cusps...14

14. Perianth segments 4–5 mm long; inner segments acute; inner filaments with median cusp half to two thirds as long as lateral cusps. SW France, Corsica, NE Spain, Balearics..**10. polyanthum**

Perianth segments 3.5–4 mm long; inner segments obtuse; inner filaments with median cusp a third to half as long as the basal lamina. Ethiopia........**31.alibile**

15. Umbel consisting of bulbils and flowers (sometimes the latter aborted)...........16

Umbel consisting wholly of flowers 18

16. Perianth segments 5–8 mm long, sparsely scabrid on keel; inner filaments with median cusp half as long as basal lamina...................................**47. dregeanum**

Perianth segments (1–)3–4(–5) mm long, smooth; inner filaments with median cusp at least twice as long as the basal lamina 17

17. Spathe up to 12 cm long; anthers subexserted; inner filaments with median cusp twice as long as the basal lamina ....**19. longicuspis**

Spathe up to 25 cm long; anthers included or equalling segments; inner filaments with median cusp ca. three times as long as the basal lamina........**19a. sativum**

18. Perianth segments (5.5–)6–9.5 mm long, acuminate; leaves 10–20 mm wide ....**12. pyrenaicum**
    Perianth segments 3–8 mm long, acute, subacute or obtuse; leaves 1–9 mm wide 19
19. Perianth segments 3–3.5 mm long, smooth on keel but with ciliate margin; umbel c.1 cm diam. NE Turkey..**23. rollovii**
    Perianth segments 3.5–8 mm long, scabrid, papillose or verrucose, at least on keel & sometimes also on margins; umbel normally at least 1.5 cm diam. 20
20. Pedicels densely papillose or pustulose, at least at the apex; outer segments densely scabrid, verrucose or coarsely papillose all over exterior; leaves 1–3 21
    Pedicels smooth or sometimes slightly papillose at apex; outer segments usually scabrid only on keel but sometimes finely papillose all over; leaves 2–4 22
21. Leaves 1–3 mm wide; segments 5–6 mm ....**42. asperiflorum**
    Leaves c.5 mm wide; segments c.4 mm ....**36. wendelboi**
22. Filaments glabrous; anthers brownish...**45. calyptratum**
    Filaments ciliate [usually minutely] in lower half; anthers yellow, or purplish before dehiscence...23
23. Perianth segments 3.5–4.5 mm long 24
    Perianth segments 5–8 mm long 26
24. Bulblets present, dark purple; outer segments lanceolate, acute or subacute. Crimea, N.Caucasus ....**33. albiflorum**
    Bulblets yellowish–brown or not produced; outer segments oblong or elliptic–ovate, subacute or obtuse...25
25. Outer bulb tunics coriaceous, brown. Transcaucasia, NE Turkey ....**25. gramineum**
    Outer bulb tunic membranous, becoming fibrous with age, greyish. Ethiopia, W Saudi Arabia..**31. alibile**
26. Leaves glabrous; outer segments 5–8 mm long. S Africa ....**47. dregeanum**
    Leaves scabrid on margins and veins; outer segments 5 mm long 27
27. Bulblets purple; perianth segments subacute, coarsely papillose-scabrid; median cusp of inner filaments two thirds as long as lateral cusps ....**35. subnotabile**
    Bulblets yellow; perianth segments acute, scabrid; median cusp of inner filaments half as long as lateral cusps ....**24. talyschense**

KEY TO GROUP D

1. Anthers partially to fully exserted 2
   Anthers included [tips of anthers may equal segments]..6
2. Perianth segments pink or purple, often with a darker median vein, usually smooth; anthers pink or violet, rarely yellow 3
   Perianth segments white, usually scabrid or pustulose; anthers yellow, pink or purple 4
3. Filaments of inner stamens with median cusp slightly shorter than or subequal to the lateral cusps; leaves 1–2(–3) mm wide (to 5 mm in cultivation)................... ....**15. cappadocicum**
   Filaments of inner stamens with median cusp half as long as lateral cusps; leaves 2–5 mm wide..**14. macrochaetum**
4. Anthers pink or purple; leaves 2–5 mm wide, smooth or scabrid on the margin; sheaths smooth; segments white with a green median vein, scabrid only on the keel ....**14. macrochaetum**
   Anthers yellow; leaves 2–3 mm wide, scabrid at least on margins; sheaths papillose-scabrid; segments white, papillose-scabrid outside 5
5. Perianth segments 2–3 mm long ....**28. pustulosum**

Perianth segments 3–3.5 mm long...**29. qaradagense**
6. Leaves glabrous...7
    Leaves scabrid or retrorse-pilose on margins...8
7. Plant 30–90 cm tall; flowers 3–5 mm long; umbel 3–6 cm diam., spherical..........
    ....**48. baeticum**
    Plant 8–24 cm tall; flowers 5.5–7 mm long; umbel 1–2.5 cm diam., hemispherical
    ....
    ....**52. gomphrenoides**
8. Plant 50–70 cm tall; leaves 4–6, about 3 mm wide; outer perianth segments
    scabrid on keel..**50. dictyoscordum**
    Plant 16–30 cm tall; leaves 2–3, 0.5–3 mm wide; outer segments smooth or slight-
    ly scabrid on margins..9
9. Leaves retrorse-pilose; spathe persistent; umbel 1–2 cm diam.; flowers white, 3–4
    mm long ....**49. longicollum**
    Leaves scabrid; spathe caducous; umbel 3–4.5 cm diam.; flowers pinkish, 5 mm
    long ....**51. gypsodictyum**

KEY TO GROUP E

1. Umbel consisting of bulbils and flowers mixed...2
    Umbel consisting wholly of flowers 5
2. Spathe falling before or at anthesis, 1-valved 3
    Spathe persistent, with 2–4 valves 4
3. Leaves 2–3; filaments glabrous; anthers violet. Iran ....**86. subvineale**
    Leaves (3–)4–5(–7) ....**85. vineale**
4. Anthers included, yellow; perianth segments acute. Portugal........**69. pruinatum**
    Anthers exserted, purple or reddish; perianth segments obtuse or subacute.
    Widespread..**62. sphaerocephalon**
5. Spathe falling at or before anthesis, usually 1-valved but the valve occasionally
    splitting [in *chamaespathum* the spathe sometimes trapped by the upper leaf and
    prevented from falling] 6
    Spathe persistent, at least until late anthesis, with (1–)2–4 valves 17
6. Anthers included; leaves sheathing only at base of stem. C Asia.........................
    ....**79. ferganicum**
    Anthers at least partially exserted at anthesis; leaves sheathing the lower fifth or
    more of the stem 7
7. Leaves sheathing the stem up to the umbel. Greece, Albania...............................
    ....**84. chamaespathum**
    Leaves sheathing stem to at most three quarters...8
8. Perianth segments scabrid or verrucose, at least on the keel of the outer three....9
    Perianth segments smooth, minutely papillose or minutely tuberculate.............10
9. Flowers purple in upper half, paler below with a darker median vein; stem 20–40
    cm. C & S Turkey...**87. stearnianum**
    Flowers white with a darker vein; stem 12–25 cm. N Egypt ....**78. mareoticum**
10. Leaves minutely scabrid on margins, sometimes only sparsely so 11
    Leaves glabrous..13
11. Bracteoles filiform, up to 2.5 cm long (but sometimes minute: see note under
    A.affine); beak of spathe up to 13 cm long; umbel spherical..............**82. affine**
    Bracteoles forming a silvery-membranous involucre or laciniate into filiform divi-
    sions to 1 cm long; beak of spathe up to 8 cm long; umbel often with a 2-tiered
    appearance at late flowering stage 12
12. Flowers purple; inner segments 3.5–6 mm long; leaves 2–8 mm wide ....

....**81. amethystinum**
Flowers white with a green median stripe or blotch, or pink or purple but if so then
inner segments 2–4 mm long; leaves 1–3 mm wide ....**80. guttatum**
13. Umbel often with a 2-tiered appearance by late flowering stage; median cusp of
inner filaments a third to half as long as the lateral cusps ....**80. guttatum**
Umbel spherical or nearly so; median cusp of inner filaments half to nearly as long
as the lateral cusps 14
14. Filaments glabrous. Iran...**86. subvineale**
Filaments [at least the outer ones] ciliate at base 15
15. Outer perianth segments minutely tuberculate on keel; bulb tunics coriaceous,
reticulate veined, ultimately breaking into weakly reticulate fibres. C Asia.......
....**108. filidentiforme**
Outer perianth segments smooth; bulb tunics membranous, the outer often split-
ting into parallel strips 16
16. Perianth segments white, c.4 mm long, obscurely keeled, acute or subacute,
glossy. C -S Turkey...**83. gorumsense**
Perianth segments pink, purple, reddish or greenish- white, 2–3.5(–5) mm long,
the outer strongly concave, subacute or obtuse. Widespread ....**85. vineale**
17. Anthers included 18
Anthers fully exserted at anthesis, although sometimes only slightly so............26
18. Perianth segments white with a purple or reddish median vein........................19
Perianth segments purple, reddish-purple, pink or violet-blue, sometimes darker-
veined..21
19. Leaves densely scabrid-papillose on veins. Lebanon, Syria ....**75. makmelianum**
Leaves glabrous. Iran, C Asia 20
20. Stem tuberculate; perianth segments 5–6.5 mm long; median cusp of inner fila-
ments a sixth as long as the basal lamina. Iran................................**76. kotschyi**
Stem smooth; perianth segments 6–7 mm long; median cusp of inner filaments a
third as long as the basal lamina. C Asia ....**77. lehmannianum**
21. Perianth violet-blue or blue. NE Turkey, Iran, Transcaucasus.............**58. aucheri**
Perianth pink to purple 22
22. Perianth segments 8–11 mm long, subequal, smooth. N Greece ....**55. heldreichii**
Perianth segments 3.5–7.5(–10) mm long, minutely papillose to scabrid, at least
on the keel; if more than 7.5 mm then inner markedly longer than outer 23
23. Inner perianth segments truncate and toothed at apex, 6.5–7.5 mm long and
markedly longer than the outer. Bulgaria, NW Turkey...................**57. jubatum**
Inner perianth segments acute or obtuse, entire, subequal, 3.5–7 mm long 24
24. Leaves 3–5 mm wide; umbel hemispherical or ovoid; anthers yellow. Portugal ...
....**69. pruinatum**
Leaves 0.5–2 mm wide; umbel with fastigiate appearance; anthers purple [?rarely
yellow in *junceum*] 25
25. Perianth 6–7 mm long; median cusp of inner filaments a sixth to a quarter as long
as the basal lamina. Cyprus, S Turkey...**54. junceum**
Perianth 3.5–4 mm long; median cusp of inner filaments a third as long as basal
lamina. Crete ....**53. rubrovittatum**
26. Perianth segments white with a greenish vein, green, greenish-yellow, green with
white margins, rarely green with a tinge of purple 27
Perianth segments pink to purple, often with a darker median vein 35
27. Perianth green with a purplish suffusion. E Turkey, Transcaucasus.......................
....**66. fuscoviolaceum**
Perianth segments white, green with white margins, green or greenish-yellow .....
....28

28. Inner filaments with the median cusp two thirds as long as to subequal to the basal lamina 29
    Inner filaments with the median cusp a third to half as long as the basal lamina ...
   31
29. Perianth pale green; spathe usually splitting into several lobes 0.5–1.3 cm long ...
    ....**60. phanerantherum**
    Perianth segments white with a straw, yellow or pale green median vein; spathe usually splitting into 2(rarely 4) lobes 1–2 cm long 30
30. Umbel 4–5 cm diam.; leaves prominently scabrid on ribs; anthers yellow. Syria ..
    ....**61. damascenum**
    Umbel usually 2–3.5 cm diam.; leaves glabrous or slightly scabrid; anthers purple. Widespread...**62. sphaerocephalon**
31. Umbel conical or ovoid-fastigiate; perianth segments green with white margins. S.Turkey, south to Egypt, Cyprus ....**67. curtum**
    Umbel usually spherical or hemispherical; perianth segments white with a green median vein, green or greenish-yellow..32
32. Perianth segments green or greenish-yellow...33
    Perianth segments white with a green median vein 34
33. Pedicels scabrid-papillose at apex; umbel 2.5–5 cm diam.; perianth segments oblong, acute. NE Turkey ....**74. artvinense**
    Pedicels smooth or very slightly papillose at apex; umbel 1.5–3 cm diam.; perianth segments oblong-ovate or oblong-obovate, obtuse. C Turkey....................
    ....**59. nevsehirense**
34. Median cusp of inner filaments a third as long as the basal lamina; outer perianth segments oblong, acute. NE Turkey, Armenia.............................**74. artvinense**
    Median cusp of inner filaments c. half as long as basal lamina; perianth segments narrowly ovate or lanceolate, obtuse or subacute. Europe, Mediterranean area
    ....**62. sphaerocephalon**
35. Outer perianth segments minutely papillose to scabrid on the outside, at least on the keel..36
    Outer perianth segments smooth..45
36. Leaves scabrid-papillose on the margins, sometimes only sparsely so...............37
    Leaves glabrous..41
37. Stem (5–)10–13(–15) cm, flexuose; leaves 2, exceeding the inflorescence. W & SW Turkey..**73. reuterianum**
    Stem 30–140 cm, straight; leaves 3–5, shorter than the inflorescence 38
38. Filaments of the inner stamens with the median cusp a third to half as long as the basal lamina; anthers only slightly exserted from perianth; umbel 1.5–2.5 cm diam. N Greece ....**63. integerrimum**
    Filaments of the inner stamens with the median cusp half as long as to subequal to the basal lamina; anthers clearly exserted; umbel usually 2–6.5 cm diam. 39
39. Perianth segments 2.5–4 mm, subacute; umbel usually 3–6.5 cm diam.; leaves 3–7 mm wide; plant up to 140 cm in height. S & W Turkey ....**64. proponticum**
    Perianth segments 3.5–6 mm, subacute, obtuse or rounded; umbel usually 2–3.5 cm diam.; leaves 1–4 mm wide, plant 15–90 cm in height 40
40. Outer stamens distinctly longer than perianth segments; median cusp of inner filaments slightly shorter to slightly longer than the lateral cusps; perianth segments obtuse or subacute. Widespread ....**62. sphaerocephalon**
    Outer stamens equalling perianth segments; median cusp of inner filaments about half as long as the lateral cusps; perianth segments obtuse to rounded. W, SW & C Turkey ....**71. stylosum**
41. Perianth broadly ovoid, dark- or blackish-purple, segments 2.5–3.5 mm long. SE

Spain ....**70. melananthum**

Perianth campanulate, ellipsoid-campanulate or cylindric-ovoid, pink to purple, segments 3.5–9.5 mm long 42

42. Umbel hemispherical-fastigiate, 0.5–2 cm diam.; anthers only partially exserted. Crete ....**53. rubrovittatum**

Umbel spherical, 1.5–3.5(–6) cm diam.; anthers exserted 43

43. Inner perianth segments subequal in length to outer, obtuse or subacute; outer segments papillose. Europe, Mediterranean region ....**62. sphaerocephalon**

Inner perianth segments noticeably longer than outer, truncate or emarginate; outer segments scabrid or verrucose-scabrid on keel 44

44. Perianth 6.5–9.5 mm long; leaves 2–3 mm wide. N Turkey (Kastamonu)............ ....**56. ilgazense**

Perianth 4–5 mm long; leaves 1–2 mm wide; N Turkey (Çankiri Prov.)............... ....**72. eldivanense**

45. Umbel conical or ovoid-fastigiate; segments purple with white margins; plant usually 15–30 cm in height. S Turkey, south to Egypt, Cyprus ....**67. curtum**

Umbel usually spherical or hemispherical; segments purple or pink, often with a darker median vein; plant usually 20–100 cm in height..................................46

46. Umbel with numerous bracteoles; filaments densely ciliate in at least the lower half; perianth segments dull purple with a green suffusion. E Turkey, Transcaucasus ....**66. fuscoviolaceum**

Umbel with few or no bracteoles; filaments glabrous, or slightly ciliate at base; perianth segments reddish-purple or pink with a darker median vein.............47

47. Outer perianth segments oblong-elliptic; median cusp of inner filaments varying from 1/2 as long as to slightly shorter than lateral cusps; pedicels papillose at apex. S Turkey, Iran, Iraq to Israel ....**60. phanerantherum**

Outer perianth segments ovate or lanceolate; median cusp of inner filaments slightly shorter or slightly longer than lateral cusps; pedicels smooth or minutely papillose..48

48. Spathe silvery-transparent, splitting irregularly into 2–4 lobes; median cusp of inner filaments 1/3 to 1/2 as long as basal lamina......................**68. regelianum**

Spathe brownish-scarious, usually splitting more or less regularly into 2 lobes; median cusp of inner filaments 1/2 as long as to subequal to basal lamina ...49

49. Umbel lax, hemispherical; perianth campanulate, 4–4.5 mm long. Balearic Is...... ....**65. ebusitanum**

Umbel dense, spherical; perianth ovoid or ovoid- cylindric, 3.5–6 mm long. Widespread Europe and Mediterranean region.................**62.sphaerocephalon**

KEY TO GROUP F

1. Anthers wholly included at anthesis 2

Anthers partially to wholly exserted at anthesis 10

2. Perianth white or white with a faint tinge of bluish- green, usually with a darker median vein...3

Perianth blue, violet-blue, purple or pink, sometimes with a darker median vein.. 8

3. Spathe falling before or at anthesis, 1-valved...4

Spathe persistent, with (1–)2–4 valves..6

4. Leaves exceeding the inflorescence; stem 10–20 cm tall; leaf sheaths with wing-like undulate ribs. NE Turkey ....**114. sosnowskyanum**

Leaves shorter than inflorescence; stem 20–60 cm tall; leaf sheaths not prominently ribbed. C Asia 5

5. Perianth segments white with a reddish median vein; leaves 2; umbel 2.5–3.5 cm diam. ....**113. crystallinum**
   Perianth segments white or very faintly tinged pink or bluish-green with a green median vein; leaves 3–4; umbel 3–6.5 cm diam. ....**107. filidens**
6. Perianth segments smooth; stems often 2–3 from each bulb, flexuose just above ground level; leaves 3–8; pedicels upward-curving. C Asia ...**100. borszczowii**
   Perianth segments scabrid-papillose, at least on the keel of the outer three; stems solitary; leaves 2–3; pedicels mostly straight 7
7. Stem 6–10(–15) cm; perianth segments c.6–7 mm long; leaves 2–2.5 mm wide. S Israel, NE Egypt, Jordan, Saudi Arabia ....**98. sinaiticum**
   Stem c.15 cm; perianth segments c.5 mm long; leaves 0.5–1 mm wide. Libya .....
   ....**93. barthianum**
8. Perianth segments violet-blue, 6–8 mm long. Israel, Jordan, Syria.......................
   ....**88. hierochuntinum**
   Perianth segments pale bluish-violet, pink or purple with a darker median vein, 4.5–5 mm long 9
9. Leaves minutely papillose on the margins, less than 1 mm wide; stem 10–25 cm. Afghanistan..**99. hedgei**
   Leaves smooth, 1.5–2 mm wide; stem c.30 cm. Turkey.......................**92. vuralii**
10. Perianth segments acuminate, pale yellowish with a brown median vein. C Asia .
    ....**112. valentinae**
    Perianth segments acute to rounded, sometimes apiculate or emarginate, white, green, brown, purple or blue, or combinations of these...................................11
11. Outer perianth segments papillose, scabrid, tuberculate or aculeolate, at least on the keel..12
    Outer perianth segments smooth...24
12. Perianth segments purplish-pink. Iraq..**95. hamrinense**
    Perianth segments white, green, brown or bluish-violet 13
13. Plant (30–)50–100 cm in height; leaves 2–5 mm wide [if less than 50 cm, umbel with caducous spathe]..14
    Plant 6–30(–40) cm; leaves 0.5–2.5 mm wide [if more than 30 cm, umbel with persistent spathe valves] 17
14. Perianth segments 5–6.5 mm long, brownish or greenish with narrow white margins. NE Turkey...**104. baytopiorum**
    Perianth segments 4–4.5 mm long, white, pale bluish-lilac or greenish. C Asia....
   15
15. Perianth segments acute, white with a green median vein...**111. brevidentiforme**
    Perianth segments obtuse, pale bluish-lilac or greenish....................................16
16. Perianth pale bluish-lilac; umbel 2–4 cm diam. ....**109. margaritiferum**
    Perianth greenish; umbel c.5 cm diam....................................**108. filidentiforme**
17. Plant 6–10(–15) cm tall; perianth 6–7 mm long; umbel 2–4 cm diam. S.Israel, NE Egypt, Jordan, Saudi Arabia ....**98. sinaiticum**
    Plant 10–30(–40) cm tall; perianth 3.5–6 mm long; umbel usually 0.8–3 cm diam. [if perianth 6 mm long, umbel only 1–1.5 cm diam.].....................................18
18. Perianth white, often with reddish or green median vein..................................19
    Perianth pale to deep bluish–violet with a darker vein 22
19. Outer perianth segments obtuse or emarginate, the keel provided with long, branched scabrid outgrowths; S.Turkey (Antalya Prov.)............**97. anatolicum**
    Outer perianth segments acute or subacute, scabrid or aculeolate on the keel ...20
20. Bulb tunic rather weakly reticulate-fibrous, developed into a long neck at apex. Israel, Jordan, Saudi Arabia ....**91. artemisietorum**
    Bulb tunic very strongly reticulate, not developed into a neck but ending in aris-

tate fibres 21
21. Perianth segments 5–6 mm long. S Turkey(Mardin province).....**90. armerioides**
 Perianth segments 4–5 mm long. C & S Turkey.........................**89. scabriflorum**
22. Leaves 1.5–2 mm wide; outer perianth segments papillose; plant c.30 cm. C -S
 Turkey...**92. vuralii**
 Leaves 0.5–1.5 mm wide; outer perianth segments scabrid to aculeolate on keel,
 although sometimes only slightly so; plant 10–30 cm.....................................23
23. Median cusp of inner filaments a third to half as long as the basal lamina; leaves
 sheathing stem for a quarter to a third. C & C -S Turkey........**89. scabriflorum**
 Median cusp of inner filaments two thirds as long as the basal lamina; leaves
 sheathing stem for about half (rarely two thirds) its length. Lebanon &
 Mt.Hermon..**94. sannineum**
24. Perianth segments green, brown or purplish, sometimes with narrow white mar-
 gins or a whitish apex 25
 Perianth segments white or very pale purple, or white tinged with bluish-green, or
 bluish-violet, often with a narrow green or purplish median stripe.................26
25. Leaves 3–11 mm wide; perianth segments 2.5–3.5 mm long; median cusp of inner
 filaments slightly shorter to slightly longer than the basal lamina ....
 ....**101. dictyoprasum**
 Leaves c.2 mm wide; perianth segments 2–2.5 mm long; median cusp of inner filaments half as long as basal lamina ....**102. karyeteinii**
26. Perianth segments 2–2.5 mm long; leaves sheathing stem for about three quarters
 of its length; median cusp of inner filaments 1.5–2.5 times longer than basal
 lamina. S Turkey(Muğla Prov.) ....**103. fethiyense**
 Perianth segments 3–6 mm long; leaves sheathing stem for a fifth to half its length
 (if to two thirds, then flowers bluish-violet); median cusp of inner filaments
 shorter than to 1.5 times as long as the basal lamina.....................................27
27. Leaves 5–15 mm wide, glabrous; robust plant 50–200 cm tall with scape to 1.7 cm
 diam. S Turkey ....**105. robertianum**
 Leaves 0.5–6 mm wide, minutely scabrid on margins, or glabrous; plant 10–100
 cm, with rather slender scape..28
28. Perianth segments retuse; anthers yellow. Desert plant from Iraq ....
 ....**96. deserti-syriaci**
 Perianth segments acute, subacute or apiculate; anthers purple or brownish. C.
 Asia...29
29. Flowers bluish-violet; plant 10–20 cm tall. Lebanon, N Israel ......**94. sannineum**
 Flowers white or whitish slightly tinged with pink or bluish-green; plant 20–100
 cm tall. C Asia 30
30. Median cusp of inner filaments half to two thirds as long as the lateral cusps.......
 ....**107. filidens**
 Median cusp of inner filaments equal to or longer than the lateral cusps...........31
31. Perianth segments 5–6 mm long; leaves 4–5, 2–6 mm wide; plant 30–100 cm;
 median cusp of inner filaments equal to or slightly longer than lateral cusps.....
 ....**106. turcomanicum**
 Perianth segments 3–4 mm long; leaves 2–3, 1–3 mm wide; plant 20–30 cm;
 median cusp of inner filaments one and a half times longer than lateral cusps...
 ....**110. brevidens**

1. Allium ampeloprasum *L.*, Sp. Pl. 1:294(1753); Sowerby & Smith, Engl. Bot. 24:t.1657(1807); Ker-Gawler in Curtis's Bot. Mag. 34:t.1385(1811); Sibth. & Smith., Fl. Graeca 4:11,t.312(1823); Reichenb., Ic. Fl. Germ. 10:t.489(1848); Syme, Engl. Bot. 9:t.1530(1869); Regel, All. Monogr. 53(1875), excl. var.; Boiss., Fl. Or. 5:232(1882); Fiori & Paol., Fl. Anal. Ital. 1:195(1896); Coste, Fl. France 3:339(1906); Aschers. & Graebn., Syn. Mitteleur. Fl. 3:105(1905); Vved. in Fl. URSS 4:253(1935), Engl. ed. 4:196(1968); Jones & Mann, Onions and their Allies 38(1963); Mouterde, Nouv. Fl. Lib. et Syr. 1:263(1966); Kollmann in Israel Journ. Bot. 20:271(1971); Kollmann in Caryologia 25:195(1972); Wendelbo in Rechinger, Fl. Iranica 76:59(1971); Bothmer in Opera Bot. Lund 34:21(1974); Bothmer in Mitt. Bot. Staatss. München 12:268(1975); Wilde-Duyfjes, Revis. Allium in Africa 63(1977), pro parte; Stearn in Ann. Mus. Goulandris 4:172(1978) & in Fl. Europaea 5:63(1980); Pastor & Valdes, Revis. Gen. Allium Penins. Iber. & Isl. Bal. 30(1983); Kollmann in Davis, Fl. of Turkey 8:163(1984); Meikle, Fl. of Cyprus 2:1617(1985); Wendelbo in Townsend & Guest, Fl. of Iraq 8:157(1985); Kollmann in Feinbrun, Fl. Palaestina 4:88(1986). Type: England, Steep Holm Is., *Newton*(lectotype BM: Herb. Sloan 152,folio 153. See Wilde-Duyfjes in Taxon 22:59(1973)).

Selected synonymy:

A.lineare Miller, Gard. Dict. ed.8,no.4(1768). Based on *A.Holmense sphaerico capite* Ray, Cat. Pl. Angl. App. (1670), from Steep Holm Is.

A.multiflorum Desf., Fl. Atlant. 1:288(1798). Type: Algeria, *Desfontaines* (holotype P).

A.porrum var. *ampeloprasum* (L.)Mirbel in Dict. Sc. Nat. 1:384(1816).

A.halleri G.Don, Monogr. All. 15(1827). Type: 'Hab. in Graecia, Helvetia, &c.' G.Don does cite a specimen, but refers to Haller, Helvet. 2:104,no.1218(1742); this in turn refers to an illustration in Micheli, Nov. Pl. Gen. t.24,no.5(1729) which could be taken as the type of *A.halleri*, and which clearly represents *A.ampelopra-sum*.

A.mogadorense Willd. in Schultes & Schultes f., Syst. Veg. 7:1004(1829). Type: Morocco, near Mogador, *Broussonet* (holotype B: Herb. Willdenow).

A.ascendens Ten., Syll. 164(1831). Type: Italy, 'Habitat in arvis Apuliae', *Tenore* (holotype NAP).

?*A.byzantinum* C.Koch in Linnaea 22:240(1849). Type: Turkey, Bosphorus, *C.Koch* (not traced).

A.gasparrini Guss., Enum. Pl. Inarime 337,t.16(1854). Type: Italy, Is. of Ischia, *Gussone* (holotype FI).

A.albescens Guss., Enum. Pl. Inarime 338(1854). Type: Italy, Is. of Ischia, *Gussone* (holotype FI).

A.ampeloprasum var. *wiedemannii* Regel, All. Monogr. 55(1875). Type: Turkey, 'Anatolia', *Wiedemann* (holotype B); the specimen labelled 'prope Angara in mont. calc. Elma daghi', 1834, *Wiedemann* (P!) is probably an isotype.

?*A.thessalum* Boiss., Fl. Or. 5:232(1882), nom. nud.

?*A.syriacum* Boiss., Fl. Or. 5:232(1882), nom. nud.

A.getulum Batt. & Trabut in Bull. Soc. Bot. France 39:75(1892). Type: Algeria, Djebel Mzi, *Battandier* (holotype MPU).

A.tortifolium Batt. & Trabut in Bull. Soc. Bot. France 39:338(1892). Type: Algeria, Ain M'lila, *Battandier* (holotype MPU).

A.ampeloprasum var. *holmense* Aschers. & Graebn., Syn. Mitteleur. Fl. 3:105(1905). Type: England, Steep Holm Is., near Bristol (no specimen cited).

A.ampeloprasum var. *gracilis* Cavara in Bull. Orto. Bot. Reg. Univ. Nap. 9:44(1928). Type: Libya, Cyrenaica, Uadi Derna, *Cavara* (not traced).

A.ampeloprasum subsp. *eu-ampeloprasum* Hayek, Prodr. Fl. Penins. Balcan. 3:39(1932).

A.ampeloprasum var. *combazianum* Maire in Bull. Soc. Hist. Nat. Afr. Nord. 25:230(1935). Type: Algeria, near Chateauneuf, *Combaz* (holotype MPU: Herb. Maire).

A.ampeloprasum var. *caudatum* Pampanini in Arch. Bot. Ital. 12:21(1936). Type: Libya, Cyrenaica, *Pampanini* 1267 (lectotype FI).

A.porrum subsp. *eu-ampeloprasum* (Hayek)Breistroffer in Bull. Soc. Sc. Dauphine 61:610(1947).

ILLUSTRATIONS. Sowerby & Smith, Engl. Bot. 24:t.1657(1807); Reichenb., Ic. Fl. Germ. 10:t.489(1848); Jones & Mann, Onions & their Allies pl.6(1963); Israel Journ. Bot. 20:265,f.1(1971); Opera Bot. Lund 34:12,f.2A-H;13,f.3H(1974); Wilde-Duyfjes, Revis. Allium in Africa 67,68,f.10,11(1976); Pastor & Valdes, Revis. Gen. Allium Penins. Iber. & Isl. Bal. 31,f.2(1983). Plate 1A-C.

Bulb ovoid or subglobose, 2–6 cm diam.; outer tunics membranous, brownish; bulblets usually present and numerous, subglobose or helmet-shaped, sometimes with an acute or shortly acuminate apex, yellowish or pale brown, produced beneath the bulb tunics. Stem 45–200 cm, stout. Leaves 4–10(–13), shorter than the inflorescence, sheathing the lower third to half of the stem, linear, non-fistulose, keeled, (3–)5–20(–40) mm wide, scabrid on the margin and keel, glaucous; sheaths glabrous. Spathe 1-valved, ovate at base, abruptly narrowed to a long beak up to about 10(–13) cm long, caducous. Umbel normally with flowers only, but rarely with flowers and bulbils mixed, spherical, (3–)5–8(–9) cm diam., dense. Pedicels unequal, up to 4.5 cm long, smooth; bracteoles present, silvery-white, laciniate at the apex. Perianth broadly campanulate or subspherical; segments white, pink or deeper reddish-pink, sometimes with a darker green or purple median vein, (3.5–)4–5.5 mm long, with large sparse papillae on the outer surface, especially on the keel, the outer ones oblong-lanceolate, elliptic or elliptic-obovate, subacute, shortly mucronate, the inner ones ovate, spathulate or elliptic, obtuse, rounded or rarely truncate, equalling or shorter than the outer. Stamens with anthers shortly exserted or sometimes equalling the segments; filaments white or purplish, strongly arching outwards, usually ciliate at the base, the outer ones simple and narrowly triangular or with an oblong base narrowed to a triangular apex, rarely with minute lateral cusps, the inner ones with the anther-bearing cusp a third to half as long as the very widely expanded undivided basal part and about half as long as the lateral cusps; lateral cusps much exceeding the segments; anthers yellow or purplish-red. Style usually exserted. Capsule ovoid or subglobose, 2.3–3.5(–4) mm long; seeds black, about 2–3.5 mm long. 2n=16,24,32,40,48,56,80.

FLOWERING TIME. April–July.

ECOLOGY. Usually in fields and former areas of cultivation, vineyards and roadsides, more rarely on rocky hillsides, cliffs and coastal beaches and in pine forests, sea level to c.2000 m.

DISTRIBUTION. Widespread in S & W Europe and N Africa, especially in Mediterranean regions, Azores, Canary Is., Turkey, Cyprus, Syria, Lebanon, Israel, Egypt, Jordan, Saudi Arabia, Iraq, Caucasus. It has been recorded in other areas, eg. SW England, Wales and Ireland, but these populations may well be the result of ancient introductions.

NOTES. *A.ampeloprasum* has the ability to persist by both copious seed production and by vegetative means since its bulbs produce many small offsets (bulblets); some variants also produce bulbils in the inflorescence; the latter bulbilliferous variants have been described as var. *babingtonii* (Borrer)Syme, Engl. Bot. 9:204(1869)[syn.

A.babingtonii Borrer] and var. *bulbiferum* Syme, op.cit.(1869)[syn. var. *bulbilliferum* Lloyd]. The former, from Cornwall, Scilly Is., W.Ireland & the Is. of Aran, has bulbils 8–15 mm long, while in the latter, which is recorded in the Channel Is. and Ile d'Yeu*, N.France, they are 6–8 mm long. Concerning the Irish taxon, Webb & Scannell (1983) consider that 'it is at least arguable that it is an endemic species [*A.babingtonii*] of NW Europe which has become to some extent commensal with man.' The species is also recorded in Wales and Anglesey (Roberts & Day 1987) but these populations, and some of those from Cornwall, consist of non-bulbilliferous plants and are therefore referable to var. *ampeloprasum*.

The very wide distribution of 'normal' (ie. non-bulbilliferous) *A.ampeloprasum* is undoubtedly due, at least in part as mentioned above, to the production of many offset bulblets around the parent bulb which are readily distributed by the agricultural activities of man. As a weed it has been recorded in Central Asia, North America, Australia and South Africa.

Although *A.ampeloprasum* appears to be most closely related to *A.commutatum* and *A.bourgeaui* it can be distinguished fairly readily from these by the rather large sparse papillae on the keel of the perianth segments; in the other two species the segments are uniformly and minutely papillose all over the outer surface. The many small, usually helmet-shaped, bulblets of *A.ampeloprasum* are also a distinctive feature. *A.atroviolaceum* is also related but has bulb tunics which split into fibres and a somewhat elongated fibrous neck to the bulb, and the flowers are a deep purple colour whereas those of *A.ampeloprasum* are usually white or pinkish and more rarely a deeper reddish shade. *A.ampeloprasum* var. *melitense* Somm. & Car.-Gatto, from Malta ("at Boschetto") is described as a much smaller plant than is normal for *A.ampeloprasum,* with the scape less than 25 cm, narrow leaves, and an umbel up to 3 cm in diameter (Borg 1927). *A.ampeloprasum* var. *hemisphaericum* Sommier ex Fiori [*A.hemisphaericum* (Sommier ex Fiori)S.Brullo; *A.ampeloprasum* subsp. *hemisphaericum* (Sommier ex Fiori)Zangheri] from the Is. of Lampedusa is described as having a hemispherical umbel; it requires further investigation in the field.

A.ampeloprasum is in need of a critical study throughout its whole range in the same very thorough way in which Bothmer (1974) has dealt with the Aegean representatives and Kollmann (1971) with those from Israel. In view of the very wide distribution this would be a lengthy task involving a great deal of field work.

A.ampeloprasum is thought to have given rise to some important crop plants, notably the leek and kurrat which are, however, rather distinct from the wild species; they are dealt with below under the names *A.porrum* and *A.kurrat*. A third variant, the 'Great Headed Garlic' resembles *A.ampeloprasum* more closely and is cultivated for its bulbs, rather than leaves as in the leek and kurrat. The large bulbs produce many offsets, as in *A.ampeloprasum*, and it is these, and sometimes the fewer but larger main 'cloves' of the bulbs, which are used for culinary purposes as an alternative to garlic although they are not as strongly flavoured.

1a. A.porrum *L.*, Sp. Pl. 1:295(1753). Type: Illustration of 'Porrum' in Dodoens, Pemptades 688(1616) (lectotype: see Wilde-Duyfjes in Taxon 22:77(1973)).

A.ampeloprasum var. *porrum* (L.)J.Gay in Ann. Sci. Nat. ser.3,8:218(1847).

ILLUSTRATIONS. Jones & Mann, Onions & their Allies pl.7a(1963)[as *A.ampeloprasum*(Leek group)]. Plate 1D.

---

*Interestingly, although generally acknowledged to be a variant of *A.ampeloprasum*, the representatives from the Ile d'Yeu have been shown to be chemically more closely allied to Garlic (*A.sativum*) than to the Leek (Boscher & Auger 1991).

Description as for *A.ampeloprasum* but bulb poorly developed; bulblets few. Leaves up to 5 cm wide, with well developed fleshy sheaths forming a false stem (ie.the portion which is normally used for culinary purposes). Umbel non-bulbilliferous, up to 20 cm diam. Perianth white or pale pink. 2n=32,48.

FLOWERING TIME. June–July.

NOTES. The leek (poireau) is commonly cultivated, especially in northern Europe, as a food plant and it is generally accepted that it was derived from *A.ampeloprasum*, probably many centuries ago; there are now many cultivars, all of which have the same basic characteristics as *A.ampeloprasum* although the bulb development is rather suppressed and it is the fleshy 'pseudostem' formed by the leaf sheaths which forms the edible part of the plant. *A.porrum* is not an annual, as stated by Fl. URSS 4:254(1935) and, if left in the soil instead of being harvested, it will produce an inflorescence and seeds and also perennate vegetatively by means of offset bulblets.

The exact means by which the Leek was derived from its wild ancestor will probably never be known; centuries of cultivation have resulted in a taxon which is distinct and, since it is such an important crop, it is considered more convenient to retain the name *A.porrum* although taxonomically and nomenclaturally this is unorthodox. There are, however, other instances, for example in the case of several cereal crop plants, where this course of action has been followed.

More information about the leek and its history can be found in Jones & Mann (1963) and in Tackholm & Drar (1954).

1b. A.kurrat *Schweinf. ex Krause* in Notizbl. Bot. Gart. Berlin 9:524(1926). Type: Egypt, *Schweinfurth* (holotype B!). Illustrations. Jones & Mann, Onions & their Allies pl.7b(1963)[as *A.ampeloprasum*(Kurrat group)]. Plate 1E.

Very similar to *A.porrum* but generally a smaller plant with narrower leaves. 2n=32.

NOTES. The kurrat is, like *A.porrum*, known only as a cultivated plant. It is grown, mainly in Egypt, Arabia, Israel and Yemen, for its leaves which are eaten raw or used for flavouring, unlike the leek in which it is normally the fleshy white leaf bases which are used. As with *A.porrum* it has been decided to retain the binomial *A.kurrat* for practical purposes.

2. Allium commutatum *Guss.*, Enum. Pl. Inar. 339(1834); Vivant in Bull. Soc. Bot. France 121:28(1975); Garbari & Cela Renzoni in Lav. Soc. Ital. Biogeogr. N.S. 5:7(1976); Zahar. in Ann. Mus. Goulandris 3:84(1977); Bothmer in Opera Bot. Lund 34:24(1974); Stearn in Ann. Mus. Goulandris 4:176(1978) & in Fl. Europaea 5:64(1980); Pastor & Valdes, Revis. Gen. Allium Penins. Iber. & Isl. Bal. 45(1983); Kollmann in Davis, Fl. of Turkey 8:166(1984). Type: Italy, Ischia, isole di S.Anna, 1 July 1853 *Gussone*(lectotype NAP).

A.wildii Heldr. in Atti Congr. Bot. Firenze 1875:232(1876); Kollmann in Israel Journ. Bot. 20:269(1971). Type: Greece, 'in insula Prasu ad Euboeam septentrionalem', July 1866, *Wild in Heldreich* 3627 (lectotype K!).
A.ampeloprasum var. *pruinosum* Boiss., Fl. Or. 5:233(1882). Type: Greece, 'in scopulo Arpedoni insularum Pharmacusarum Atticae', *Heldreich* (holotype G?).
A.rotundum subsp. *commutatum* (Guss.)Nyman, Consp. Fl. Europ. 735(1882); K.Richter, Pl. Eur. 1:201(1890).
A.ampeloprasum var. *lussinense* Haračić in Verh. Zool. Bot. Ges. Wien 43:46(1894). Type: former Yugoslavia, 'Skoglio Karabus bei Lussin', 27 July 1893, *Haračić*

(lectotype K!, isolectotype DR, JE, M).

?*A.pruinosum* Candargy in Bull. Soc. Bot. France 44:140(1897). Type: Greece, Is. of Lesvos, Perama, *Candargy* (not traced).

A.aestivalis J.J.Rodr., Fl. Men. 180(1904). Type: Balearic Is., Menorca, 'Binisaida, Cala Mezquita, penascos al oeste de Mahon' (type not traced).

A.polyanthum var. *aestivalis* (J.J.Rodr.)Pau in Butl. Inst. Catalana Hist. Nat. 14:140(1914).

A.bimetrale Gand., Fl. Cret. 99(1916). Type: Crete, Kissamos, Gonia, 90 m, 24 June 1916, *Gandoger* 4880 (lectotype LY, isolectotype K!).

A.ampeloprasum var. *commutatum* (Guss.)Fiori, Nuova Fl. Anal. Italia 266(1923).

A.scopulicolum Font Quer in Butl. Inst. Catalana Hist. Nat. 24:144(1924). Type: Balearic Is., Ibiza, 'in scopulo Espardell del Frare de l'Espartar, dicto, et alibi, in saxosis calcareis maritimus Ebussi' (type not traced).

A.ampeloprasum subsp. *bimetrale* (Gand.)Hayek, Prodr. Fl. Penins. Balcan. 3:40(1932).

A.porrum subsp. *bimetrale* (Gand.)Breistroffer in Bull. Soc. Sc. Dauphine 6:610(1947).

A.ampeloprasum subsp. *commutatum* (Guss.)Zangheri, Fl. Ital. 1:861(1976).

ILLUSTRATIONS. Israel Journ. Bot. 20:269,f.5(1971)[as *A.wildii*]; Opera Bot. Lund 34:28,f.10(1974); Ann. Mus. Goulandris 3:84,f.4–13(1977). Plate 2A–B.

Bulb ovoid, up to about 6 cm diam.; outer tunics membranous or coriaceous; bulblets present, rather large, (0.6–)1–3(–4.4) cm long, ovoid, acute or shortly acuminate, yellowish or brown, produced beneath the bulb tunics. Stem 50–100(–180) cm. Leaves 5–13(–15), shorter than the inflorescence, sheathing the lower third to half of the stem, linear, non-fistulose, keeled, 9–25(–45) mm wide, scabrid on the margin and keel; sheaths glabrous. Spathe 1-valved, ovate at base, abruptly narrowed to a long beak up to about 30 cm long, caducous. Umbel spherical or hemispherical, 3–7(–7.5) cm diam., dense. Pedicels up to 3 cm long, smooth, swollen at the base; bracteoles present, silvery-white. Perianth ovoid or subspherical; segments whitish-green, yellowish-green, pinkish or purplish, 3–5(–5.5) mm long, densely and minutely papillose all over the outer surface, the outer broadly elliptic to ovate, obtuse or acute, inner spathulate, obtuse, truncate or emarginate. Stamens with anthers well-exserted; filaments minutely scabrid-ciliate at the base, the outer ones usually tricuspidate, rarely simple or with two cusps, the inner ones tricuspidate or sometimes with 5 cusps, the anther-bearing cusp two thirds to three quarters as long as the expanded undivided basal part and about three fifths as long as the lateral cusps; lateral cusps much exceeding the segments but usually much-contorted; anthers yellow or reddish. Style well-exserted. Capsule ovoid or subglobose, 2.5–4 mm long; seeds black, about 2.5–4 mm long. 2n=16,24,32,48.

FLOWERING TIME. May–June.

ECOLOGY. Usually to be found in open rocky or stony places near the sea on islands, sea level to about 300 m.

DISTRIBUTION. Aegean Is., Cyclades, W.Crete, Ionian Is., Peloponnese, Adriatic coastal islands of former Yugoslavia, Sicily, Ischia, Corsica, Sardinia, Balearic Is. Rarely recorded on the mainland of Greece, W.Turkey and coastal former Yugoslavia.

NOTES. *A.commutatum* is closely related to *A.ampeloprasum* and *A.bourgeaui*. It can be distinguished from both of these on account of the tricuspidate outer filaments; from *A.ampeloprasum* it also differs in having much larger bulblets, a longer spathe, and the perianth segments are uniformly and minutely papillose all over the outer surface. The shape of the inner perianth segments may be used to distinguish further between *A.commutatum* and *A.bourgeaui*; in the former they are wider towards the apex, spathulate or obovate, while in the latter they are oblong.

3. Allium bourgeaui *Rech. fil.* in Ann. Naturh. Mus. Wien. 47 :150(1936) & Fl. Aegaea 715(1943); Bothmer in Opera Bot. Lund 34:22(1974); Stearn in Ann. Mus. Goulandris 4:175(1978) & in Fl. Europaea 5:64(1980); Kollmann in Davis, Fl. of Turkey 8:164(1984).

Bulb ovoid, about 1.5 cm diam.; outer tunics membranous; bulblets present, rather large, up to 1.5 cm long, compressed-ovoid, shortly acuminate, brown or yellowish, produced beneath the bulb tunics. Stem 45–115 cm. Leaves 4–11, shorter than the inflorescence, sheathing the lower third to half of the stem, linear, non-fistulose, keeled, 6–25 mm wide, scabrid on the margin; sheaths glabrous. Spathe 1-valved, ovate at base, abruptly narrowed to a beak up to 15 cm long, caducous. Umbel spherical or hemispherical, 3–6 cm diam., dense. Pedicels 1–2.5 cm long, smooth or slightly scabrid, swollen at the base; bracteoles absent or, if present, few, subulate and inconspicuous, silvery-white. Perianth campanulate-cylindrical; segments purple, pink, reddish or pale green, 2.5–5 mm long, the outer ones lanceolate, acute or obtuse, keeled, papillose or scabrid-papillose, especially along the keel, the inner ones narrowly oblong, obtuse or truncate. Stamens with anthers well-exserted; filaments ciliate at the base, the outer simple, narrowly triangular, the inner ones with the anther-bearing cusp half as long as the expanded undivided basal part and between a quarter and a third as long as the lateral cusps; lateral cusps much exceeding the segments; anthers yellow, reddish or purple. Style well-exserted. Capsule ovoid-globose, about 4 mm long; seeds black, about 3–4 mm long.

NOTES. *A.bourgeaui* is closely related to *A.ampeloprasum* and *A.commutatum*. It is distinguished from the former by having fewer, larger bulblets, and the perianth segments papillose all over the outer surface, and from the latter by its oblong inner perianth segments and simple outer filaments.

Three subspecies are recognised:

1. Perianth segments pale green ....subsp. **bourgeaui**
   Perianth segments purple, reddish or pink 2
2. Outer segments 4–5 mm long, with small and large papillae on the outside; flowers reddish or pink ....subsp. **creticum**
   Outer segments 2.5–4 mm long, only with small papillae on the outside; flowers purple ....subsp. **cycladicum**

3a subsp. **bourgeaui**. Type: Greece, Karpathos, near Phiniki, 18 June 1935, *Rechinger* 8300(isolectotypes BM,K!,LD!,W).

ILLUSTRATIONS. Ann. Naturh. Mus. Wien 47:150,f.1f–h(1936); Opera Bot. Lund 34:13,f.3i(1974); Fl. of Turkey 8:165,f.6,no.2(1984).

Flowers campanulate or cylindrical, pale green or whitish-green; outer segments 2.5–4 mm long, with uniform small papillae on the outside; anthers yellow. 2n=16,32.

FLOWERING TIME. June–July.
ECOLOGY. Limestone cliffs and rocky hillsides, 50–800 m.
DISTRIBUTION. E.Aegean Is., Rhodes, Karpathos, Kasos; SW Turkey (Burdur province).

3b subsp. **cycladicum** *Bothmer* in Opera Bot. Lund 34:23,41(1974); Zahar. in Ann. Mus. Goulandris 3:86(1977); Stearn in Ann. Mus. Goulandris 4:176(1978) & in Fl. Europaea 5:64(1980); Kollmann in Davis, Fl. of Turkey 8:166(1984). Type: Greece, Cyclades, Is. of Paros, 2–2.5 km SE of Leukas, 250 m, 30 June 1964, *Bothmer & Strid* B.75(holotype LD!).

Illustrations. Opera Bot. Lund 34:23,f.2j-k(1974); Ann. Mus. Goulandris 4:176,24(1978).

Flowers campanulate or cylindrical, purple; outer segments 2.5–4 mm long, with uniform small papillae on the outside; anthers purple. 2n=16,24,32.

Flowering time. June–August.
Ecology. Rocky hillsides, in scrub and on cliffs, 30–250 m.
Distribution. Greece, E.Peloponnese, Cyclades, Ikaria; SW Turkey(Muğla & Denizli provinces).

3c subsp. **creticum** *Bothmer* in Mitt. Bot. Staatss. München 12:272(1975); Stearn in Ann. Mus. Goulandris 4:176(1978) & in Fl. Europaea 5:64(1980). Type: Crete, Ep. Sitia, 1 km SSW of Tourloti, 200 m, 22 May 1974, *Bothmer* B.857(holotype LD!).

Flowers ellipsoid or ovoid, reddish or pinkish; outer segments 4–5 mm long, with small and large papillae on the outside, especially on the keel; anthers yellow, red or rarely purple. 2n=16,32.

Flowering time. May–July.
Ecology. Limestone cliffs, up to 250 m.
Distribution. Crete.

4. Allium pseudoampeloprasum *Miscz. ex Grossh.*, Fl. Kavk. ed.1,1:204(1928); Kollmann in Davis, Fl. of Turkey 8:167(1984). Type: USSR, Armenia, Shorbulag, Erivan, *Grossheim* (isotype LE!, photo K!).

A.germainae Grossh. in sched.

Illustrations. Fl. Kavk. ed.2,2:t.14,f.6,6a(1940); Fl. of Turkey 8:165,f.6,no.5 (1984) [but see note below].

Bulb ovoid, 1–2 cm diam.; outer tunics greyish-black, membranous; bulblets present, yellowish. Stem 40–60 cm, smooth. Leaves 4–5, sheathing about the lower third to half of the stem, linear, canaliculate, non-fistulose, (4–)6–8 mm wide, glabrous. Spathe 1-valved, broadly ovate at the base, narrowing abruptly to a long beak exceeding the umbel, caducous. Umbel spherical, (3.5–)4–5(–5.5) cm diam., dense. Pedicels 0.7–2 cm long; bracteoles present, few, minute, white. Perianth ovoid or campanulate; segments pale pink or whitish with a darker purple or green midvein, 3.5–5 mm long, shiny, acute, the outer lanceolate, scabrid on the outside, or sometimes only on the keel and margins, the inner slightly shorter, ovate or oblong-lanceolate, scabrid. Stamens exserted; filaments ciliate at the base, the outer ones simple, triangular-subulate, the inner ones with the antherbearing cusp subequal to the undivided basal part and two thirds to three quarters as long as the lateral cusps; lateral cusps long-exserted from the perianth; anthers pale violet before dehiscence. Style well-exserted. Capsule subglobose, 2–3 mm long; seeds black, about 2–2.5 mm long. 2n=16.

Flowering time. June–July.

ECOLOGY. Dry clayey, stony or rocky slopes, 1200–2500 m.

DISTRIBUTION. NE Turkey (recorded in Erzurum, Maraş & Van provinces); Armenia.

NOTES. Very similar to *A.ampeloprasum* but differing in having the central cusp of the inner filaments about equal in length to the basal lamina; in *A.ampeloprasum* the median cusp is only about a half to two thirds as long as the basal lamina; furthermore, the perianth segments of *A.pseudoampeloprasum* are covered with small papillae whereas in *A.ampeloprasum* they are either smooth or have large sparse papillae along the keel. N.B. The illustration in Fl. of Turkey 8:f.6,no.5(1984) is incorrect since the median cusp of the inner filaments is shown to be very much shorter than the basal lamina.

5. Allium iranicum *(Wendelbo)Wendelbo* in Townsend & Guest, Fl. of Iraq 8:158(1985). Type: Iran, Aq Bolagh-e Aqdaq, 100 km N. of Hamadan, *Rioux & Golvan* 34 (holotype W).

A.ampeloprasum subsp. *iranicum* Wendelbo in Rechinger, Fl. Iranica 76:58(1971).

ILLUSTRATIONS. Fl. Iranica 76:t.6,f.82b(1971). Plate 2C.

Bulb ovoid, up to 3 cm diam.; outer tunics greyish-brown, membranous at first, sometimes splitting longitudinally but not becoming markedly fibrous; bulblets present, yellowish-brown, long-stipitate, situated beneath the leaf sheaths well above the parent bulb and usually above soil level. Stem 60–80 cm. Leaves 4–6, shorter than the inflorescence, sheathing the lower third to half of the stem, linear, shallowly canaliculate, non-fistulose, keeled, 8–10 mm wide, glabrous; sheaths glabrous. Spathe 1-valved, up to 8 cm long, broadly ovate narrowed to a long beak, caducous. Umbel spherical, 5–5.5 cm diam., dense. Pedicels up to 2 cm long, smooth; bracteoles present. Perianth campanulate; segments pale pinkish-purple, 3.5–4.5 mm long, acute or subacute, the outer narrowly ovate, keeled, scabrid, the inner ovate, sparsely scabrid. Stamens with anthers exserted; filaments glabrous, the outer simple, triangular-subulate, the inner ones with the anther-bearing cusp subequal to, or about three quarters as long as, the widely expanded undivided basal part and a half to two thirds as long as the lateral cusps; lateral cusps well-exserted; anthers yellow. Style exserted. Capsule subglobose, about 3 mm long.

FLOWERING TIME. June–July.

ECOLOGY. Stony slopes in scrub, thorny steppe, 1500–2600 m.

DISTRIBUTION. Iran, NE Iraq.

NOTES. Closely related to *A.ampeloprasum* but differs on account of the long-stipitate bulblets borne on slender stolons beneath the leaf sheaths, by having acute or subacute inner and outer perianth segments and by the glabrous filaments.

6. Allium leucanthum *C.Koch* in Linnaea 22:240(1849); Vved. in Fl. URSS 4:252(1935), Engl. ed. 4:195(1968); Grossh., Fl. Kavk. ed.2,2:120(1940); Kollmann in Israel Journ. Bot. 20,4:272(1971). Type: Transcaucasia, Shirvan steppe, 50–200 m, *C.Koch* [not traced, probably destroyed; neotype chosen from type locality, *Kolakovsky* (LE)!].

A.ampeloprasum var. *leucanthum* (C.Koch)Ledeb., Fl. Ross. 4:164(1852); Regel, All. Monogr. 54(1875); Boiss., Fl. Or. 5:232(1884).

A.firmotunicatum var. *album* Grossh., Fl. Kavk. ed.1,1:204(1928). Type: Armenia,

Erivan area.

A.ampeloprasum subsp. *leucanthum* (C.Koch)Hayek, Prodr. Fl. Penins. Balcan. 3:40(1932).

A.firmotunicatum forma *album* (Grossh.)Grossh., Fl. Kavk. ed.2,2:120(1940)

ILLUSTRATIONS. Israel Journ. Bot. 20,4:268,f.6;269,f.4(1971).

Bulb ovoid-globose, 2–3 cm diam.; outer tunics greyish-brown, membranous, splitting and becoming fibrous with age; bulblets present, numerous, yellowish-brown. Stem 50–120 cm. Leaves 4–7, shorter than the inflorescence, sheathing the lower quarter to third of the stem, linear, shallowly canaliculate, non-fistulose, keeled, 3–9 mm wide, scabrid; sheaths glabrous. Spathe 1-valved, caducous. Umbel spherical, 3.5–6 (–7) cm diam., dense. Pedicels up to 2.8 cm long, smooth; bracteoles present. Perianth campanulate-ellipsoid; segments white with a green median vein, 3–3.5 mm long, the outer narrowly oblong-ovate, acute or subacute, keeled, smooth or very slightly scabrid along the keel, the inner elliptic-obovate, obtuse or rounded, smooth. Stamens with anthers exserted; filaments ciliate, the outer simple, triangular-subulate, the inner ones with the anther-bearing cusp two thirds as long as the expanded undivided basal part and two thirds to three quarters as long as the lateral cusps; lateral cusps well-exserted; anthers yellow. Style exserted. Capsule subglobose, about 3–3.5 mm long. 2n=16,32.

FLOWERING TIME. June–August.

ECOLOGY. Dry places and in fields.

DISTRIBUTION. E & S Transcaucasus; N Iran; E Saudi Arabia.

NOTES. Closely related to *A.ampeloprasum*. It has small white flowers with a green median vein on each of the narrow perianth segment, rather smaller bulbs with fibrous tunics, a generally more slender habit and narrower leaves. White-flowered variants of *A.ampeloprasum* from other regions outside the Caucasus have been erroneously identified as *A.leucanthum*. It should be said, however, that the species has not been thoroughly studied in the wild and further work is required to establish whether the distinguishing characteristics are reliable.

The record for Saudi Arabia is based on *Dickson* 184A(K!), collected at Kaba near the border with Kuwait.

7. Allium talijevii *Klokov* in Kotov & Barbarich, Fl. RSS Ucr. 3:113,406(1950). Type: Ukraine, SE of Donetsk, near Maschlykovka by the Krynka river, 1898, *V. I. Talijev* (holotype KW).

ILLUSTRATIONS. Klokov, loc. cit.:115,f.12(1950).

Bulb ovoid, 2.5–3 cm diam.; outer tunics yellowish-white or yellowish-brown, membranous, becoming finely reticulate-fibrous at the apex; bulblets numerous, yellowish, plano-convex, ovate, 5–6 mm long. Stem 45–50 cm. Leaves 5–7, sheathing the lower three quarters of the stem, linear, flat, non-fistulose, 5–12 mm wide, scabrid on the margin and on the veins beneath; upper leaf much-exceeding the umbel. Spathe caducous. Umbel hemispherical, 2.5–4 cm diam. Pedicels 1.2–2.5 mm long; bracteoles present. Perianth ovoid-campanulate; segments yellowish-white, about 3.5 mm long; outer oblong-lanceolate with a smooth or slightly scabrid keel, inner oblong-elliptic. Stamens exserted; filaments glabrous, the outer ones simple, lanceolate-subulate, the inner ones with the anther-bearing cusp shorter than the lateral cusps; anther colour not known. Capsule not known.

FLOWERING TIME. June.

ECOLOGY. Chalky slopes.

DISTRIBUTION. SE Ukraine.

NOTES. Related to *A.leucanthum* but having glabrous filaments and longer leaves.

8. Allium pardoi *Loscos*, Trat. Pl. Arag. 1:9(1876); Willk., Ill. Fl. Hisp. 2:2(1886); Maire, Fl. Afr. Nord. 5:259(1958); Stearn in Ann. Mus. Goulandris 4:177(1978) & in Fl. Europaea 5:65(1980); Pastor & Valdes, Revis. Gen. Allium Penins. Iber. & Isl. Bal. 43(1983). Type: Spain, Castelseras, *Loscos* (not traced).

ILLUSTRATIONS. Ill. Fl. Hisp. 2:t.96(1886); Revis. Gen. Allium Penins. Iber. & Isl. Bal. 44,f.5(1983).

Bulb ovoid, 2.5–4 cm diam.; outer tunics greyish, membranous, eventually splitting into parallel fibres; bulblets numerous, boat-shaped, acute, yellowish. Stem 80-110 cm, the apex coiled or nodding before anthesis. Leaves 3–7, shorter than the inflorescence, sheathing the lower quarter to half of the stem, linear, shallowly canaliculate, non-fistulose, keeled, 7–10 mm wide, minutely denticulate on the margin; sheaths glabrous. Spathe 1-valved, exceeding the inflorescence, 4.5–9 cm long, ovate at the base, tapering to an acuminate-beaked apex, usually caducous. Umbel spherical or ovoid, 3–5 cm diam., dense. Pedicels 1–3 cm long, smooth; bracteoles present, membranous, somewhat laciniate. Perianth narrowly ovoid-campanulate; segments white with a green mid-vein, smooth, the outer lanceolate, subacute, keeled, 4–4.2 mm long, the inner ovate, obtuse, sometimes shortly and bluntly acuminate, 2.8–3.5 mm long. Stamens exserted; filaments ciliate near the base, the outer ones simple, narrowly triangular, the inner ones with the anther-bearing cusp about two thirds as long as the expanded undivided basal part and slightly shorter than, to about two thirds as long as, the lateral cusps; lateral cusps exserted; anthers brownish-yellow. Style exserted. Capsule subglobose, about 3 mm long; seeds black, about 3 mm long. 2n=48.

FLOWERING TIME. June–July.

ECOLOGY. Near fields, sea level–400 m.

DISTRIBUTION. NE Spain, in the provinces of Teruel and Zaragoza.

NOTES. This is undoubtedly very close to *A.ampeloprasum* and was included in that species by Wilde-Duyfjes (1976); Stearn (op. cit.) and Pastor & Valdes (op. cit.), however, maintained it at specific rank. It may be distinguished from *A.ampeloprasum* on the basis of the smooth perianth segments and slightly smaller flowers, and by the fact that the central cusp of the inner filaments is shorter in relation to the lateral cusps than in the latter species.

9. Allium scaberrimum *Serres* in Bull. Soc. Bot. France 4:439(1857); Coste, Fl. France 3:338(1906); Stearn in Ann. Mus. Goulandris 4:175(1978) & in Fl. Europaea 5:64(1980). Type: France, 'dans les bles de la pleine de La Roche pres Gap', *Serres* (specimen not traced).

A.rotundum var. *scaberrimum* (Serres)Ascher. & Graebn., Syn. Mitteleur. Fl. 3:104(1905).

ILLUSTRATIONS. Coste, Fl. France 3:338,fig.(1886).

Bulb ovoid; outer tunics pale brownish, membranous, sometimes splitting lengthways; bulblets ovoid, acuminate, produced beneath the tunics of the parent bulb, yellowish-brown. Stem 40–80 cm. Leaves about 4, shorter than the inflorescence, sheathing the lower third to half of the stem, linear, canaliculate, non-fistulose, keeled, 7–12 mm wide, conspicuously dentate on the margins; sheaths scabrid on the veins. Spathe 1-valved, about 3 cm long, ovate at the base, tapering to a beaked apex, caducous. Umbel spherical or hemispherical, 3–5 cm diam., dense. Pedicels up to 2.5 cm long, smooth; bracteoles present, up to about 4 mm long, white-membranous, somewhat laciniate. Perianth ovoid; segments whitish, 3–4.5 mm long, lanceolate or oblong, the outer subacute, sparsely papillose-warted, especially on the keel, the inner obtuse or rounded, smooth. Stamens with anthers slightly exserted; filaments ciliate near the base, the outer ones simple, triangular-subulate, the inner ones with the anther-bearing cusp two thirds as long as or subequal to the expanded undivided basal part and about half as long as the lateral cusps; anthers yellow. Style exserted. Capsule subglobose, about 3 mm long.

FLOWERING TIME. June–July.
ECOLOGY. Fields, approx. 1000–1500 m.
DISTRIBUTION. S France, Hautes-Alpes, near Gap; NW Italy, Piedmont, in the Cuneo area.
NOTES. Although very similar to *A.ampeloprasum* this is rather different in that the outer surface of the outer perianth segments is furnished with sparse low papillae and, in the inner filaments, the ratio of the length of the median cusp to the basal lamina is distinct in the two species. It is, however, a plant which requires further study and assessment.

10. Allium polyanthum *Schultes & Schultes fil.*, Syst. Veg. 7:1016(1830); Coste, Fl. France 3:338(1906); Rouy, Fl. France 12:249(1910); Stearn in Ann. Mus. Goulandris 4:175(1978) & in Fl. Europaea 5:64(1980); Pastor & Valdes, Revis. Gen. Allium Penins. Iber. & Isl. Bal. 36(1983). Types: France, 'in arenosis et cultis Niceae, Telonis, Narbonae, agri Ruscinonencis Tolosae', no collectors stated (not traced).

A.multiflorum DC. in Lam. & DC., Fl. Franc. 3rd. ed. 6:316(1815), non Desf.
A.ampeloprasum subsp. *polyanthum* (Schultes & Schultes fil.)O.de Bolos, J.Vigo,
    R.M.Masalles & J.M.Ninot, Fl. Mar. Paisos Catalans 1213(1990).

ILLUSTRATIONS. Coste, Fl. France 3:338,fig.(1886); Revis. Gen. Allium Penins. Iber. & Isl. Bal. 37,f.3(1983).

Bulb ovoid, 1–2 cm diam.; outer tunics pale brownish, membranous; bulblets helmet-shaped or hemispherical, produced around the base of the parent bulb, yellowish. Stem (15–)20–50(–80) cm. Leaves 2–3(–6), shorter than the inflorescence, sheathing the lower quarter, rarely up to a third, of the stem, linear, shallowly canaliculate, non-fistulose, keeled, (2–)4–10(–20) mm wide, smooth or occasionally minutely denticulate on the margin; sheaths glabrous. Spathe 1-valved, shorter than the inflorescence, 1.5–3 cm long, ovate at the base, tapering to an acuminate apex, usually caducous. Umbel spherical or hemispherical, 2–5(–8) cm diam., dense. Pedicels 0.5–2(–3) cm long, smooth; bracteoles present, small, white-membranous, somewhat laciniate. Perianth cylindrical-ovoid; segments pale pink with a darker pink mid-vein, 4–5 mm long, elliptic or oblong, the outer acute or obtuse and bluntly mucronate, keeled, papillose, the inner often slightly wider, acute, smooth or papillose. Stamens with anthers included; filaments ciliate near the base, the outer ones simple, narrowly triangular, the inner ones with the anther-bearing cusp about a third as long as the expanded undivided basal part and a half to two thirds as long as the lateral cusps; lateral cusps contorted and included or slightly exserted; anthers yellowish. Style exserted. Capsule subglobose, about 3.5 mm long; seeds black, about 3–3.5 mm long.

2n=32.

FLOWERING TIME. May–July.

ECOLOGY. In and near fields, sea level–400 m.

DISTRIBUTION. SW France; Corsica; NE Spain; Balearic Is., recorded on Mallorca, Ibiza, Cabrera, & Formentera.

NOTES. Closely related to *A.ampeloprasum* and included in that species by Wilde-Duyfjes (1976) but maintained at species level by Stearn (op. cit.) and Pastor & Valdes (op. cit.). It is usually a smaller plant than *A.ampeloprasum* with shorter perianth segments; the generally fewer leaves, which are often smooth-margined, sheath the stem for about a quarter to a third of its length whereas in *A.ampeloprasum* they are always scabrid-margined and they sheath the stem for a third to a half.

11. Allium acutiflorum *Loisel.* in Desv., Journ. Bot. Redige 2:279(1809); Reichenb., Ic. Fl. Germ. 10:23(1848); Coste, Fl. France 3:338(1906); Greuter in Bull. Soc. Bot. France 121:146(1974); Stearn in Ann. Mus. Goulandris 4:178(1978)& in Fl. Europaea 5:65(1980). Type: France, 'dans le Piemont, à Tende et au Mont Gros', *M.Perret* (holotype ?P).

ILLUSTRATIONS. Ic. Fl. Germ. 10:t.491(1848); Bicknell, Fl. Pl. Riviera t.78(1885).

Bulb ovoid, 1–1.5 cm diam.; outer tunics pale brown, membranous or subcoriaceous; bulblets apparently absent. Stem 15–50 cm, glabrous. Leaves 2–4, shorter than the inflorescence, sheathing the lower fifth to third of the stem, usually still green at flowering time, linear-lanceolate, tapering to a long-acuminate apex, flat or shallowly canaliculate, non-fistulose, keeled, 1–6 mm wide, minutely denticulate on the margin; sheaths glabrous. Spathe 1- or 2-valved, up to 3 cm long, membranous or subcoriaceous, ovate at the base, tapering to a short beak, often persisting until anthesis. Umbel hemispherical, (2–)2.5–5(–6) cm diam., dense. Pedicels up to 2 cm long, smooth; bracteoles present, membranous, 3–5 mm long, dissected at the apex. Perianth campanulate; segments pale purplish-pink with a darker mid-vein, 6–9 mm long, lanceolate, acuminate, slightly papillose-scabrid on the keel. Stamens included; filaments ciliate near the base, the outer ones simple, narrowly triangular, the inner ones with the anther-bearing cusp a fifth to a quarter as long as the expanded undivided basal part and a quarter to half as long as the lateral cusps; lateral cusps included; anthers yellow. Style included. Capsule ovoid-globose, about 4 mm long; seeds black, about 3 mm long. 2n=16.

FLOWERING TIME. May–June.

ECOLOGY. Sandy and rocky places, mainly at low altitudes near the sea.

DISTRIBUTION. S France, Corsica, NW Italy.

NOTES. Easily distinguished by the large pink flowers with acuminate segments and included stamens.

12. Allium pyrenaicum *Costa & Vayreda* in Costa, Introd. Fl. Cataluña, ed.2, supl.92(1877); Vayreda in An. Soc. Esp. Hist. Nat. 9:87(1880); Willk., Ill. Fl. Hisp. 1:124(1884), Suppl. Prodr. Fl. Hisp. 51(1893); Coste & Soulie in Bull. Soc. Bot. France 59:560(1912); Stearn in Ann. Mus. Goulandris 4:178(1978) & in Fl. Europaea 5:65(1980); Pastor & Valdes, Revis. Gen. Allium Penins. Iber. & Isl. Bal. 48(1983). Type: Spain, E.Pyrenees, Coll de Marrem (=Malrem), June 1872, *Vayreda* (lectotype BC/ Herb. Sennen).

A.controversum Costa, Introd. Fl. Cataluña, ed.2,supl.73(1877), non Schrader ex Willd.(1809). Type as above.

ILLUSTRATIONS. Ill. Fl. Hisp. 1:t.75(1884); Pastor & Valdes, Revis. Gen. Allium Penins. Iber. & Isl. Bal. 49,f.7(1983).

Bulb ovoid, 2–4 cm diam.; outer tunics pale brown, membranous; bulblets apparently absent. Stem 55–120 cm, glabrous. Leaves 4–7, shorter than the inflorescence, sheathing the lower quarter to third of the stem, usually still green at flowering time, linear-lanceolate, tapering to a long-acuminate apex, flat or shallowly canaliculate, non-fistulose, keeled, 1–2 cm wide, minutely denticulate on the margin; sheaths glabrous. Spathe 1-valved, up to 7 cm long, membranous, ovate at the base, tapering to an acuminate-beaked apex, usually caducous. Umbel spherical or hemispherical, 3–7(–9) cm diam., dense. Pedicels up to 2.7(–3.5) cm long, smooth; bracteoles present, membranous. Perianth cylindric-campanulate; segments white with a green midvein, unequal or subequal, lanceolate or oblong-lanceolate, acuminate, the outer (5.5–)6.5–9.5 mm long, papillose on the keel and with serrate margins, the inner 6–8 mm long, smooth or with slightly serrate margins. Stamens included; filaments ciliate near the base, the outer ones simple, narrowly triangular, the inner ones with the anther-bearing cusp about a third to half as long as the expanded undivided basal part and about half to two thirds as long as the lateral cusps; lateral cusps included; anthers brownish. Style included. Capsule subglobose, about 4–6 mm long; seeds black, about 3–5 mm long. 2n=16,32.

FLOWERING TIME. June–July.
ECOLOGY. Rocky places, 150–1300 m.
DISTRIBUTION. Spain, E.Pyrenees in the provinces of Gerona and Huesca.
NOTES. Similar to *A.acutiflorum* in that the flowers are large with acuminate perianth segments but differing in having white perianth segments which are conspicuously serrate on the margin; the structure of the inner filaments is also rather different.

13. Allium atroviolaceum *Boiss.*, Diagn. Pl. Or. Nov. 1,7:112(1846) & Fl. Or. 5:240(1882); Hayek, Prodr. Fl. Penins. Balcan. 3:40(1932); Vved. in Fl. URSS 4:252(1935), Engl. ed. 4:194(1968); Zahar. in Savulescu, Fl. Reip. Pop. Romania 11:252(1966); Kollmann in Israel Journ. Bot. 29:272(1971); Wendelbo in Rechinger, Fl. Iranica 76:58(1971); Stearn in Ann. Mus. Goulandr. 4:175(1978)& in Fl. Europaea 5:64(1980); Kollmann in Davis, Fl. of Turkey 8:168(1984); Wendelbo in Townsend & Guest, Fl. of Iraq 8:159(1985). Type: Iran, Sabz Pushan, near Shiraz, 31 May 1842, *Kotschy* 450 (holotype G, isotype P!).

A.ampeloprasum var. *atroviolaceum* (Boiss.)Regel, All. Monogr. 54(1875).
A.atroviolaceum var. *caucasicum* Somm. & Lev. in Acta Horti Petrop. 16:427(1900).
    Type: Caucasus, near Tiflis, 8 July 1890, *Sommier & Levier* (holotype FI photo!).
A.firmotunicatum Fomin in Monit. Jard. Bot. Tiflis 14:48(1909). Type: Caucasus,
    Mil'skaya steppe, *Schelkovnikov* (not traced).
A.atroviolaceum var. *firmotunicatum* (Fomin)Grossh., Fl. Kavk. ed.2,2:120(1940).
A.atroviolaceum var. *ruderale* Grossh., Fl. Kavk. ed.2,2:120(1940).

ILLUSTRATIONS. Fl. Iranica 76:t.6,f.81(1971); Israel Journ. Bot. 29:f.3(1971); Fl. of Turkey 8:165,f.6,no.6(1984). Plate 2D.

Bulb ovoid, 1–2.5 cm diam.; outer tunics greyish-brown, membranous at first, becoming parallel-fibrous with age and extended into a long neck at the apex; bulblets present, yellowish-brown. Stem 50–100(–150) cm. Leaves 3–5, shorter than the inflorescence, sheathing the lower quarter to a third of the stem, linear, shallowly canaliculate, non-fistulose, keeled, 2–10 mm wide, scabrid on the keel and margins; sheaths glabrous or scabrid. Spathe 1-valved, with a long beak, caducous. Umbel spherical, 2.5–3(–6) cm diam., dense. Pedicels up to 2.5 cm long, smooth; bracteoles present, small, white-membranous or sometimes up to 6 mm long, laciniate and conspicuous. Perianth urceolate-campanulate; segments usually deep purple-maroon (occasionally greenish forms occur in populations of purple ones), 3.5–5 mm long, the outer ovate, elliptic-ovate or oblong, subacute or obtuse, keeled, smooth or scabrid, the inner elliptic or ovate, obtuse or rounded, smooth. Stamens with anthers clearly exserted; filaments ciliate near the base, the outer simple, triangular-subulate, the inner ones with the anther-bearing cusp about two thirds as long as the expanded undivided basal part and varying from a half as long as to about equal to the lateral cusps; lateral cusps well-exserted; anthers dark purple. Style exserted. Capsule subglobose, about 2.5 mm long. 2n=16,32[24,40,48 also recorded].

FLOWERING TIME. June–July.
ECOLOGY. Fields, vineyards, dry hillsides, rocky places, sea level–2000 m.
DISTRIBUTION. Bulgaria; Czech Rep.; Slovakia; Greece; Hungary; Italy; former Yugoslavia; Roumania; Turkey; Ukraine; Crimea; Caucasia; C Asia; N Iraq; Iran; Afghanistan.
NOTES. Closely related to *A.ampeloprasum* and similar in its overall characteristics but can usually be distinguished by the bulb tunic which splits and becomes distinctly fibrous, and by the dark purple flowers.

A.firmotunicatum Fomin was said to differ from *A.atroviolaceum*, mainly in its smaller stature, narrower leaves, fewer-flowered umbels and smaller perianth, and in its preference for xeric habitats. However, these features appear to vary even within populations and do not seem to correlate with distribution or habitat. A geographically wide survey is required in order to assess the importance of bracteole size. In most specimens they are small and inconspicuous but occasionally they are considerably larger and quite obvious, eg. *Davis* 6605A (K!) from Syria.

13a. Allium sp. cf. atroviolaceum from Saudi Arabia. Representative specimen: Saudi Arabia, Thumamah, 80 km NNE of Riyadh, 610 m, 10 April 1990, *I.S.Collenette* 7425 (K!).

Bulb ovoid, 1–2.5 cm diam.; outer tunics greyish-brown, membranous at first, becoming slightly fibrous with age; bulblets sometimes present, brown. Stem usually 45–70(–100) cm. Leaves 3–5, shorter than the inflorescence, sheathing the lower quarter to a third of the stem, linear, shallowly canaliculate, non-fistulose, keeled, 2–10 mm wide, smooth or slightly scabrid on the keel and margins; sheaths glabrous. Spathe 1-valved, with a long beak, caducous. Umbel spherical, 1.5–3(–7) cm diam., dense. Pedicels up to 3 cm long, smooth; bracteoles present, white-membranous. Perianth ellipsoid-campanulate; segments usually deep purple, 3.5–5 mm long, the outer ovate, elliptic-ovate or oblong, subacute or obtuse, keeled, smooth, the inner elliptic or ovate, obtuse or rounded, smooth. Stamens with anthers slightly exserted at anthesis; filaments ciliate near the base, the outer simple, triangular-subulate, the inner ones with the anther-bearing cusp about equal in length to the expanded undivided basal part and varying from a half as long as to about equal to the lateral cusps; lateral cusps well-exserted; anthers dark purple. Style exserted. Capsule subglobose, about 2.5 mm long.

FLOWERING TIME. February-April.
ECOLOGY. Sand dunes, rock crevices, 550–2440 m.

DISTRIBUTION. Saudi Arabia; Kuwait; S Desert region of Iraq.

NOTES. The specimens cited below probably represent *A.atroviolaceum* but 'species 13a' is included here in order to draw attention to a detail which requires further study. The Arabian material collected by Mrs.I.S.Collenette has inner filaments with the median cusp about equal to the basal lamina (2/3 as long in *A.atroviolaceum*). However, in some of the individuals the median cusp appears to be slightly shorter. In all other respects the material is within the range of variation of the very widespread and often weedy *A.atroviolaceum* so, rather than describing a new taxon, it is probably more appropriate to extend the presently accepted range of variation in the filament characters within this species to include the material cited below.

Saudi Arabia: *Al-Khalili* 372; *Collenette* 2504, 5082, 6382, 6584, 6779, 7425, 7605, 7816; Kuwait: *Boulos & Cope* 17677, *Dickson* 184; Iraq: *Rawi* 14196 (all specimens at K).

14. Allium macrochaetum *Boiss. & Hausskn.* in Boiss., Fl. Or. 5:239(1882); Feinbr. in Pal. Journ. Bot., Jer. Ser. 3:7(1943); Zoh. in Dep. Agr. Iraq Bull. 31:35(1950); Rawi in Dep. Agr. Iraq. Tech. Bull. 14:185(1964); Mouterde, Nouv. Fl. Lib. et Syr. 1:267(1966); Kollmann in Davis, Fl. of Turkey 8:169(1984); Wendelbo in Townsend & Guest, Fl. of Iraq 8:156(1985); Type: Iraq, "in uliginosis salsis Assyriae prope Tell Afar et Mosul", May 1867, *Haussknecht* 971(holotype G, isotype BM!).

A.laeve Wendelbo & Bothmer in Rechinger, Fl. Iranica 76:57(1971). Type: Iran, Luristan, 30 km S of Kuh-i Dasht, *Wendelbo* 2025(holotype BG!).

ILLUSTRATIONS. Feinbr., Pal. Journ. Bot., Jer. Ser. 3:19,f.13(1943); Mouterde, Nouv. Fl. Lib. Syr. 1,Atlas:t.85,f.1(1966); Fl. of Turkey 8:165,f.6,no.8(1984); Fl. Iranica 76:t.6,f.80;t.16,f.3(1971)(as *A.laeve*). Plate 3A–B.

Bulb ovoid, 1–2(–2.5) cm diam., often purplish- or reddish-tinged; outer tunics greyish or brownish, reticulate-fibrous or sometimes almost membranous, produced into a distinct neck up to 5 cm long; bulblets few, elongate, sometimes absent. Stem 25–60(–100) m. Leaves sometimes remaining green until flowering time, sometimes withering away, 3–4, sheathing the lower third of the stem, linear, canaliculate, non-fistulose, 2–5 mm wide, margin smooth or papillose-scabrid. Spathe 1-valved, with a beak 2–6 cm long, caducous. Umbel more or less spherical or oblong, 2–4.5 cm diam., dense. Pedicels unequal, 0.5–2 cm long, swollen at the apex; bracteoles usually present, linear-filiform, white. Perianth campanulate; segments pinkish or purplish with a darker mid-vein, or white with a green mid-vein, 2.5–3.5 mm long, ovate or elliptic, obtuse or subacute, smooth or papillose-scabrid on the margins and keel. Stamens with anthers exserted; filaments smooth or very slightly papillose at base, the outer ones simple, triangular-subulate, the inner ones with the anther-bearing cusp slightly shorter than or subequal to the undivided basal part and about 1/2 as long as the very long lateral cusps; lateral cusps long-exserted from the perianth but often contorted and thus appearing to be rather shorter than they really are; anthers usually pink or purplish before dehiscence but sometimes yellow. Style slightly exserted. Capsule globose, 2.5–3 mm long; seeds black, about 2 mm long. 2n=16.

FLOWERING TIME. May–July.

ECOLOGY. Saline clay soils, sandy places, stony or rocky slopes, 100–1650 m.

DISTRIBUTION. NW Iran; N Iraq; Syria; C & SE Turkey (recorded in provinces of Ankara, Erzincan, Hakkâri, Konya, Maraş, Sivas and Van).

NOTES. This is closely allied to *A.atroviolaceum* but the outer bulb tunics are distinctly reticulate-fibrous; *A.cappadocicum* is also rather similar but this has the median cusp of the inner three segments almost equal in length to the lateral cusps, where-as in *A.macrochaetum* it is only about half as long. The bulbs of *A.macrochaetum* are generally larger than those of *A.cappadocicum* and the outer tunics are more distinct-ly fibrous. *A.macrochaetum* smells of garlic when crushed and could almost certain-ly be used as an alternative. It is also allied to *A.tuncelianum* (which also smells of garlic): see further notes under *A.tuncelianum*.

15. Allium cappadocicum *Boiss.*, Fl. Or. 5:241(1882); Kollmann in Davis, Fl. of Turkey 8:169(1984). Type: Turkey, Kayseri prov., "in collibus ad occasum Caesareae [Kayseri] in Cappadocia sitis", *Balansa* 1115 (holotype G, isotype JE!,K!,P!).

ILLUSTRATIONS. Fl. of Turkey 8:165,f.6,no.7(1984). Plate 3C.

Bulb ovoid, 1–2(–3) cm diam.; outer tunics greyish, splitting into weakly reticu-lated fibres, produced into a long neck up to 6 cm long; bulblets absent. Stem 20–30 (–45) cm. Leaves usually withered away at flowering time, 2–4, sheathing the lower third to half of the stem, linear, canaliculate, non–fistulose, 1–2(–3) mm wide (some-times to 5 mm in cultivated specimens), smooth. Spathe 1-valved, with a long beak, caducous. Umbel more or less spherical or ovoid, (1.5–)2–3(–4) cm diam., dense. Pedicels unequal, 0.5–2 cm long; bracteoles present, conspicuous, silvery-white. Perianth cylindrical or campanulate, smelling of dung; segments pinkish or purplish with a darker mid-vein, or sometimes white with a green mid-vein, shiny, 3–4 mm long, oblong-elliptic, obtuse, smooth. Stamens with the anthers partly to fully exsert-ed at anthesis; filaments glabrous, the outer ones simple, triangular-subulate, the inner ones with the anther-bearing cusp slightly shorter than the undivided basal part and only slightly shorter than or subequal to the lateral cusps; lateral cusps slightly to well exserted from the perianth; anthers violet before dehiscence. Style exserted. Capsule globose, 2.5–3 mm long; seeds black, about 2 mm long. 2n=16.

FLOWERING TIME. June–July.
ECOLOGY. Sandy places, dry stony or rocky slopes, roadsides, 900–1400 m.
DISTRIBUTION. C Turkey (recorded in Ankara, Çorum, Kayseri, Konya, Nevşehir and Yozgat provinces).
NOTES. This is closely allied to *A.atroviolaceum* but has smooth leaves, shorter pedicels and glabrous filaments. *A.macrochaetum* and *A.cappadocicum* are also closely related and comments about the differences can be found under the former species, no. 14. Further fieldwork is required in order to clarify the relationships between these since they are very similar. In the living state the flowers of *A.cap-padocicum* smell of dung, similar to those of *A.fuscoviolaceum*.

16. Allium asirense. *B.Mathew* in Kew Bull. 51(1): in press. Type: Saudi Arabia, 5 km north of Mindak, Al Bahah-Taif road, *I.S.Collenette* 8251 (holotype K!).

ILLUSTRATION. Kew Bull. 51(1): in press.

Bulb ovoid, 0.5–1.5 cm diam.; outer tunics greyish, membranous with parallel fibres; bulblets pale brown, helmet-shaped. Stem 20–30 cm. Leaves usually withered away at flowering time, 2–4, sheathing the lower fifth to a third of the stem, linear, canaliculate, non-fistulose, 1–2 mm wide, prominently scabrid on the margins. Spathe 1-valved, ovate at the base narrowed abruptly to a beak to 2.5 cm long, membranous, pinkish-tinged, early caducous. Umbel more or less spherical, 1.2–2 cm diam., very dense. Pedicels unequal, 2–9 mm long; bracteoles few, minute, white (some of these are probably aborted flowers). Perianth campanulate; segments deep purple with paler margins, shiny, 3–4 mm long, oblong-elliptic, obtuse and bluntly mucronate, smooth or very slightly warted on the keel. Stamens with included to partially exserted anthers; filaments slightly ciliate at base, the outer ones simple, triangular-subulate, the inner ones with the anther-bearing cusp half as long as the undivided very widely expanded basal part and half to two thirds as long as the lateral cusps; lateral cusps included or slightly exserted from the perianth; anthers violet before dehiscence. Style included. Capsule and seeds not seen.

FLOWERING TIME. May–June.
ECOLOGY. Shale hillsides among junipers and in fallow fields, 2134 m.
DISTRIBUTION. W.Saudi Arabia
NOTES. This is apparently related to *A.cappadocicum* Boiss. from Turkey, but the bulbs produce many helmet-shaped bulblets and the plants often grow in clumps rather than singly. The tunics of the mature bulbs consist of parallel fibres, not at all reticulate as in *A.cappadocicum*, the leaves are prominently scabrid on the margins (glabrous in *A.cappadocicum*) and there are also differences between the two species in the relative proportions of the median cusp of the inner filaments to the basal lamina and to the lateral cusps. Four collections have been seen, all from the same general area on the Al Bahah to Taif escarpment road: *I.S.Collenette* 6729, 8248 & 8251 (K) and *A.A.Fayed* 1464 (K,KTUH & Assiut University).

17. Allium sandrasicum *Kollmann, N.Özhatay & Bothmer* in Notes R.B.G. Edinb. 41:264, fig.10(1983); Kollmann in Davis, Fl.of Turkey 8:167(1984). Type: Turkey, Muğla province, Köyceğis, Sandras Dağ, 10 July 1979, *E.Özhatay* 2465(ISTE 43996) (holotype ISTE!).

ILLUSTRATIONS. Fl. of Turkey 8:165,f.6,no.4(1984). Plate 3D.

Bulb ovoid, 1–1.8 cm diam., outer tunics black, membranous; bulblets few, yellow or brownish, oblong-ovoid, fusiform, stipitate, often borne on stem. Stem 50–65 (–90) cm. Leaves 2–4, flat, linear, keeled beneath, 2.5–4(–8) mm wide, margin sometimes slightly scabrid, sheathing the lower third to half of the stem; undulate. Spathe caducous, wide at the base, narrowed to beak 5–8 cm long. Umbel globose, (1.5–)3–5 cm diam. Pedicels 1.5–2 cm long, very slender. Perianth broadly campanulate; segments white with a green mid-vein, smooth, 2–2.5 mm long, subequal in length but inner broader than outer; outer oblong, inner obovate, emarginate. Stamens with anthers strongly exserted; median cusp of inner filaments shorter than lateral cusps and somewhat shorter than basal lamina; anther colour not known. Style exserted. Capsule 3.5–4 mm long. 2n=16.

FLOWERING TIME. June–July.
ECOLOGY. Limestone cliffs, near seashore, crevices of rocky outcrops, 10–300m.
DISTRIBUTION. Turkey: Muğla and Antalya provinces.
NOTES. A slender small-flowered species, said by the original authors to be allied to *A.bourgeaui* but easily distinguished from this by the small white smooth perianth segments and narrower leaves. Further collections, including living material, are desirable in order to elucidate its relationships.

Although normally the umbel is wholly floriferous, one specimen flowering at Kew in 1991 (*N.Özhatay* ISTE 60508) produced a bulbil in the inflorescence.

The inflorescence of *A.sandrasicum* bears a close resemblance to that of *A.guttatum* subsp. *sardoum*, but the leaves are flat, not fistulose, so they are in fact not closely related.

18. Allium truncatum *(Feinbr.)Kollmann & D.Zohary* in Feinbr., Fl. Palaestina 4:90(1986); Kollmann in Rotem 15:61(1985)[without basionym]. Type: Israel, Negev, Dhahiriya to Be'er Sheva 28 May 1942, *D.Zohary & Feinbrun* (lectotype HUJ!).

A.ampeloprasum subsp. *truncatum* (Feinbr.)Kollmann in Israel Journ. Bot. 20:271(1971).
A.ampeloprasum var. *truncatum* Feinbr. in Pal. Journ. Bot., Jer. Ser. 3:6(1943).
A.ampeloprasum var. *portorii* Gombault in Bull. Soc. Bot. France 96:10(1949). Type: Syria, Khan Abou Chamat, NE of Damascus, 4 May 1934, *Gombault* 5734 (holotype P!).

ILLUSTRATIONS. Pal. Journ. Bot., Jer. Ser. 3:19,f.10(1943); Fl. Palaestina 4(Plates):pl.123(1986). Plate 4A.

Bulb ovoid or subglobose, 2–4 cm diam.; outer tunics grey, membranous, sometimes splitting lengthways; bulblets ovoid-subglobose, produced on slender stolons up to 2.5 cm long. Stem (80–)100–150 cm. Leaves several, shorter than the inflorescence, sheathing the lower third of the stem, linear, canaliculate, non-fistulose, keeled, 8–15 mm wide, with slightly scabrid or smooth margins; sheaths smooth. Spathe 1-valved, about 8 cm long, broadly ovate at the base, tapering to long beak, caducous. Umbel spherical, 3–8 cm diam., dense . Pedicels up to 3.5 cm long, smooth; bracteoles present, up to about 4 mm long, white-membranous, somewhat laciniate. Perianth campanulate; segments dark purple or rarely green, 2.5–4 mm long, the outer ones oblong-ovate, obtuse, scabrid on the outside, the inner ones oblong or obovate, truncate, emarginate or slightly dentate at the apex, slightly longer, sparsely scabrid. Stamens with anthers exserted; filaments ciliate near the base, the outer ones simple, triangular-subulate, the inner ones with the anther-bearing cusp half to two thirds as long as the expanded undivided basal part and about half as long as the contorted lateral cusps; anthers purple before dehiscence. Style exserted. Capsule subglobose. 2n=16,24,32.

FLOWERING TIME. April–June.
ECOLOGY. 'In Irano-Turanian territories in crop fields on loess (Negev), Lissan marls and basalt soils(Jordan Valley); in Mediterranean territories on alluvial and sandy soils, on rendzina derived from Eocene and Senonian calcareous rocks.'(from Flora Palaestina 4:90(1986).
DISTRIBUTION. N.Egypt, Israel, Jordan, Syria.
NOTES. *A.truncatum* is probably more closely related to *A.bourgeaui* than to *A.ampeloprasum*, although in the past it has been regarded as a subspecies of the latter. It is not difficult to distinguish from the latter since it has deep purple flowers (rarely green) with truncate inner segments, and it differs from both in having bulblets borne at the ends of long stolons. The proportions of the cusps of the inner filaments are an additional point of difference from *A.bourgeaui*, and it is frequently a taller plant with narrower leaves, although there is an overlap in these features and these statements should be regarded as generalisations applicable to populations rather than selected individuals.

19. Allium longicuspis *Regel*, All. Monogr. 45(1875); Vved. in Fl. URSS 4:243(1935), Engl. ed. 4:188(1968); Vved., Consp. Fl. Asiae Mediae 2:79(1971); Wendelbo in Rechinger, Fl. Iranica 76:53(1971); Kollmann in Davis, Fl. of Turkey 8:163(1984). Types: C Asia, 'prope Taka in Kokania', *O.Fedtschenko* (holotype LE).

ILLUSTRATIONS. Fl. Iranica 76:t.5,f.72(1971). Plate 4B.

Bulb ovoid, 1–2 cm diam.; outer tunics greyish-brown, membranous or subcoriaceous; tunics of new bulbs tinged reddish; replacement bulbs 2–4. Stem 40–110 cm, smooth, coiled at the apex prior to anthesis. Leaves 4–7, much shorter than the inflorescence, narrowly linear, flat, canaliculate, non-fistulose, 5–20 mm wide, smooth or slightly scabrid on the margin and keel, sheathing the lower third to half of the stem; sheaths smooth. Spathe 1-valved, broadly ovate at base with a long beak up to 12 cm long, caducous. Umbel approximately spherical, containing ovoid, beaked, purplish-tinged bulbils and usually rather few flowers. Pedicels unequal, up to 2.5 cm long, smooth; bracteoles present, white, scarious, ovate and acuminate to linear, 5–8 mm long. Perianth ovoid-campanulate; segments pale pinkish or nearly white, 3–4 mm long, acute, smooth, narrowly lanceolate, narrowly oblong-ovate or narrowly obovate, often with erose-dentate margins. Stamens exserted or subexserted; filaments glabrous, the outer ones simple, linear-subulate, the inner ones with the anther-bearing cusp about twice as long as the undivided basal part and a third to a half as long as the lateral cusps; lateral cusps well-exserted from the perianth; anthers yellow. Style exserted, white, 4 mm long. Capsule and seeds not seen. 2n=16.

FLOWERING TIME. July–August.

ECOLOGY. Rocky valleys and river flats, 1350–2100 m.

DISTRIBUTION. Turkey(recorded in Hakkâri province); NW Iran; C Asia (recorded in Kopet Dag, Pamir Alai and Tien Shan Mts.).

NOTES. This apparently wild species is very similar to garlic, *A.sativum*, which has presumably been derived from it. *A.longicuspis* also seems to be closely allied to the Turkish *A.tuncelianum* which has a fully fertile, non-bulbilliferous inflorescence; further comments are made under *A.tuncelianum*.

19a A.sativum *L.*, Sp. Pl. 296(1753). Type: Herb. Burser 111:90 (UPS) [lectotype: see Wilde-Duyfjes in Taxon 22:81(1973)].

ILLUSTRATIONS. Reichenb., Ic. Fl. Germ. 10:t.488,f.1069(1848); Jones & Mann, Onions & their Allies pl.5b & 12a(1963). Plates 4C-D; 5A-B.

Similar to *A.longicuspis* but with the following characteristics. Bulb depressed-globose, when mature consisting of up to 15(–60) large bulblets ('cloves'), produced beneath a common tunic. Leaves 4–12. Spathe up to 25 cm long. Flowers usually aborting before anthesis, greenish-white or pinkish; perianth segments (1–)3(–5) mm long, lanceolate, acuminate. Stamens with anthers included or equalling the segments; filaments glabrous, the outer ones simple or tricuspidate, the inner 3–, 5– or rarely 7–cuspidate, with the median cusp about three times as long as the basal lamina and a third to half as long as the lateral cusps. 2n=16[rarely 12,18].

NOTES. Garlic has been cultivated widely since ancient times for its bulbs which are much prized for flavouring. Realistic clay models of garlic bulbs, dating back to 3000 B.C. or earlier, have been found in Egypt (Täckholm & Drar, 1954) and actual bulbs from about 1500 B.C. were found in the tomb of Tutankhamen. Today there are many cultivars, which have presumably arisen by vegetative mutation ('sports') since garlic is a sterile plant, although there is the possibility that some past clones of *A.sativum* may have been capable of producing seeds and thus could have led to the breeding and selection of variants by other means. Interestingly, there is a report in the 1990 IBPGR publication *Geneflow* that collectors for CMEA, the Council for Mutual Economic Aid, have located in Central Asia four samples of garlic which do have fertile pollen. This phenomenon has also been noted in Caucasian material

(Etoh, Kojima & Matsuzoe, 1992).

A.longicuspis and *A.sativum* are very similar morphologically and it is not unlikely that the former may in fact represent an old clone of garlic which has become locally naturalized where conditions happened to be particularly favourable; this might account for the rather scattered records of *A.longicuspis* in western Asia. A similar view has been expressed by H.Nielsen [Economically important species of the genus Allium: an IBPGR report (ined.)].

A variant of garlic which has the upper part of its stem more markedly coiled before anthesis has been described as *A.sativum* var. *ophioscorodon* (Link)Döll [syn. *A.ophioscorodon* Link, *A.controversum* Schrader, *A.sativum* subsp. *ophioscorodon* (Link)J.Holub]. This is sometimes known as Serpent Garlic or Rocambole; the latter name has also been used for *A.scorodoprasum*.

A further variant which has wide leaves drooping over at their tips, has been described as *A.sativum* var. *pekinense* (Prokhanov)Makino [syn. *A.pekinense* Prokhanov]. This eastern Asiatic plant should, however, be regarded as another cultivar of garlic, not as a botanical variety.

20. Allium tuncelianum *(Kollmann) N.Özhatay, B.Mathew & Şiraneci in Kew Bull.* 51(1): in press.

A.macrochaetum Boiss. & Hausskn. subsp. *tuncelianum* Kollmann in Notes R.B.G. Edinb. 41:262(1983); Kollmann in Davis, Fl. of Turkey 8:170(1984). Type: Turkey, Tunceli province, Munzur Dağ, Aksu Dere above Ovacik, 1800 m, 21 July 1957, *Davis* 31498(holotype E, isotype K!).

ILLUSTRATIONS. Notes R.B.G. Edinb. 41:262,f.6(1983); Fl. of Turkey 8:170(1984). Plate 5C.

Bulb subglobose-ovoid, 2–4 cm diam.; outer tunics greyish-brown, membranous; bulblets present,0.8–1.5 cm long, ovoid, acuminate, pale brown, produced beneath the bulb tunics. Stem 40–100 cm, coiled in the upper part before maturity. Leaves 4–6, shorter than the inflorescence, sheathing about the lower quarter of the stem, linear, non-fistulose, keeled, 0.6–1.8 cm wide, smooth; sheaths glabrous. Spathe 1-valved, ovate at base, abruptly narrowed to a long beak 7–16 cm long, caducous. Umbel spherical, 5–8 cm diam., very dense . Pedicels up to 3.5 cm long, smooth, very slender; bracteoles absent or few and very inconspicuous. Perianth campanulate; segments pale pinkish-mauve or whitish with a darker purple or green median vein, 2.5–3 mm long, glabrous, narrowly oblong, narrowly obovate or narrowly lanceolate, obtuse or rounded-truncate. Stamens with anthers well-exserted; filaments glabrous, the outer ones simple, narrowly triangular, the inner ones with the anther-bearing cusp up to three times longer than the expanded undivided basal part (?rarely subequal to) and about one third to half as long as the very long (up to 5 mm) contorted lateral cusps; lateral cusps much exceeding the segments; anthers pinkish-purple. Style well-exserted. Capsule subglobose, about 4 mm long; seeds black, about 3.5–4 mm long. 2n=16.

FLOWERING TIME. July–August
ECOLOGY. Rocky slopes, screes, 1100–2200 m.
DISTRIBUTION. C & E Turkey (recorded in the provinces of Erzincan, Sivas, Tunceli).
NOTES. *A.tuncelianum* is of interest in that its vegetative parts smell of Garlic and it is used as such locally in Turkey. Further studies are required to ascertain the relationships between this species, *A.sativum* and *A.longicuspis* but it is apparent that they

PLATE 1

A. **A.ampeloprasum.** [Kew Acc.No.1973-1884]. Photo Media Resources, R.B.G.Kew

B. **A.ampeloprasum,** 'bulbs' and helmet-shaped 'bulblets'. Photo B.Mathew.

C. **A.ampeloprasum,** showing proliferating umbel. Scilly Isles. Photo B.Mathew

D. **A.porrum,** the Leek. Photo B.Mathew.

E. **A.kurrat,** inflorescence. Photo B.Mathew.

PLATE 2

A. **A.commutatum** [Kew Acc.No.1977-1734] Photo B.Mathew.
B. **A.commutatum** in Greece, Paxos Is. Photo O.Fragman.
C. **A.iranicum** [Kew Acc.No.1977-4268]. Photo Media Resources, R.B.G.Kew
D. **A.atroviolaceum.** Photo B.Mathew.

PLATE 3

A. **A.macrochaetum** [Kew Acc.No.1990-2462]. Photo Media Resources, R.B.G.Kew
B. 'Cloves' of bulb of **A.macrochaetum** [Kew Acc.No.1990-2462]. Photo B.Mathew.
C. **A.cappadocicum** [Kew Acc.No.1990-2049]. Photo Media Resources, R.B.G.Kew
D. **A.sandrasicum** [Kew Acc.No.1989-2740]. Photo Media Resources, R.B.G.Kew

PLATE 4

A. **A.truncatum** [G.Shuz, Israel]. Photo B.Mathew.
B. **A.longicuspis** [J.Ruksāns, C.Asia]. Photo Media Resources, R.B.G.Kew
C. Bulb of **A.sativum** [Garlic], showing 'cloves'. Photo B.Mathew.
D. **A.sativum,** inflorescence. Photo Media Resources, R.B.G.Kew

PLATE 5

A.   Garlic on sale in China. Photo P.Cribb.

B.   Garlic on sale in China. Photo J.Cowley.

C.   **A.tuncelianum** [Kew Acc.No.1986-3639]. Photo Media Resources, R.B.G.Kew

D.   **A.oltense** [Kew Acc.No.1985-2209]. Photo B.Mathew.

PLATE 6

A. **A.alibile** [Kew Acc.No.1989-1569]. Photo Media Resources, R.B.G.Kew

B. **A.scorodoprasum** in Slovenia. Photo C.C.Townsend.

C. **A.rotundum** [Baines 19]. Photo Media Resources, R.B.G.Kew

D. **A.rotundum,** bulbs [Kew Acc.No.1990-1173]. Photo B.Mathew.

PLATE 7

A. **A.rotundum** in Turkey. Photo B.Mathew.

B. **A.erubescens** [Furse 7127]. Photo Media Resources, R.B.G.Kew

C. **A.enginii** in Turkey [Kew Acc.No.1990-2204]. Photo J.Cowley.

D. **A.enginii** [Kew Acc.No.1990-2204]. Photo B.Mathew.

E. **A.enginii** habitat in Turkey. Photo J.Cowley.

PLATE 8

A. **A.dregeanum** [Kew Acc.No.1990-0801]. Photo Media Resources, R.B.G.Kew
B. **A.rubrovittatum** [Kew Acc.No.1969-1217]. Photo Media Resources, R.B.G.Kew
C. **A.gomphrenoides** [Kew Acc.No.1984-3642]. Photo B.Mathew.

PLATE 9

A. **A.junceum** [Kew Acc.No.1988-2745]. Photo Media Resources, R.B.G.Kew
B. **A.heldreichii** [Kew Acc.No.1971-4170]. Photo Media Resources, R.B.G.Kew
C. **A.jubatum** in Turkey near Bolu. Photo M.Johnson.
D. **A.phanerantherum** [Kew Acc.No.1990-1172]. Photo Media Resources, R.B.G.Kew

PLATE 10

A.   **A.jubatum** [Kew Acc.No.1990-2035]. Photo Media Resources, R.B.G.Kew

B.   **A.aucheri** [Kew Acc.No.1989-2756]. Photo Media Resources, R.B.G.Kew

C.   **A.sphaerocephalon** [Kew Acc.No.1977-3385]. Photo Media Resources, R.B.G.Kew

D.   **A.sphaerocephalon** bulb pack on sale. Photo B.Mathew.

PLATE 11

A. **A.nevsehirense** [Kew Acc.No.1990-2074]. Photo Media Resources, R.B.G.Kew

B. **A.fuscoviolaceum** [Baines & Henry 116c]. Photo Media Resources, R.B.G.Kew

C. **A.curtum** [Salmon & Lovell 37]. Photo Media Resources, R.B.G.Kew

D. **A.curtum** [Kew Acc.No.1984-0154]. Photo Media Resources, R.B.G.Kew

PLATE 12

A. **A.stylosum** [Kew Acc.No.1990-2479]. Photo Media Resources, R.B.G.Kew

B. **A.artvinense** [Baines & Henry 76]. Photo Media Resources, R.B.G.Kew

C. **A.guttatum** ssp. **guttatum** [Baines 16]. Photo Media Resources, R.B.G.Kew

D. **A.guttatum** ssp. **sardoum** [Kew Acc.No.1974-3476]. Photo Media Resources, R.B.G.Kew

PLATE 13

A. **A.amethystinum.** Photo B.Mathew.

B. **A.amethystinum** [Kew Acc.No.1984-0497]. Photo Media Resources, R.B.G.Kew

C. **A.affine** [Kew Acc.No.1990-2158]. Photo Media Resources, R.B.G.Kew

D. **A.chamaespathum** [Kew Acc.No.1989-3204]. Photo Media Resources, R.B.G.Kew

E. **A.vineale** [left Kew Acc.No.1976-4048, right 1972-0727]. Photo Media Resources, R.B.G.Kew

PLATE 14

A.   **A.stearnianum** [Kew Acc.No.1990-1653]. Photo B.Mathew.
B.   **A.stearnianum** in Turkey. Photo M.Johnson.
C.   **A.hierochuntinum** [Kew Acc.No.1988-2784]. Photo Media Resources, R.B.G.Kew
D.   **A.scabriflorum** [Kew Acc.No.1989-2727]. Photo Media Resources, R.B.G.Kew
E.   **A.scabriflorum** in Turkey. Photo J.Cowley.

PLATE 15

A. **A.artemisietorum** [Kew Acc.No.1990-1171]. Photo Media Resources, R.B.G.Kew
B. **A.dictyoprasum** [Baines & Henry 53]. Photo Media Resources, R.B.G.Kew
C. **A.dictyoprasum** in Turkey. Photo J.Cowley.

PLATE 16

A. **A.filiden**s [Stevens s.n.]
B. **A.sosnowskyanum** in Turkey. Photo M.Johnson.
C. **A.sosnowskyanum** [J.White 2-90]. Photo B.Mathew.
D. **A.turkestanicum** [Fritsch 781]. Photo Media Resources, R.B.G.Kew

have certain features in common as well as the characteristic odour, notably the coiling of the flower stem before anthesis, the pale-coloured small, glabrous, rather narrow perianth segments and the glabrous filaments with very long lateral cusps.

The possibility exists that the sterile or near-sterile bulbilliferous *A.sativum* and *A.longicuspis* could be derived from *A.tuncelianum* (or perhaps *A.macrochaetum*), and it is desirable that this conjecture is explored in greater detail than is possible in this broad survey of sect. *Allium*.

A.tuncelianum is rather distinctive on account of the very numerous flowers on filiform pedicels, which are often tinted pinkish-purple.

21. Allium pseudophanerantherum *Rech. fil.* in Ark. Bot., ser.2,1:506(1951); Mouterde, Nouv. Fl. Lib. et Syr. 1:264(1966). Type: Syria, Qaryatein, 25 April 1935, *Mouterde* 39 (holotype S).

ILLUSTRATIONS. Nouv. Fl. Lib. et Syr. 1(Atlas):pl.83,no.4(1966).

Bulb ovoid, about 3 cm diam.; tunics white, membranous, produced into a neck 8 cm long; bulblets pale brown, 0.8 cm long. Stem about 50 cm or more in height, circinnate at first, becoming erect. Leaves sheathing the lower half of the stem, linear, canaliculate, non-fistulose, 5–7 mm wide, scabrid on the margin; sheaths slightly scabrid. Spathe caducous. Umbel rather few-flowered, lax, about 5–5.5 cm diam., with many sterile flowers on short pedicels at the base (this may be an abnormal condition; see note below). Pedicels up to 2.5 cm long, glabrous; bracteoles apparently present but obscured by the sterile flowers. Perianth urceolate-ovoid, umbilicate at the base; segments pale pink, 5 mm long, ovate, attenuate at the apex, shiny. Stamens with the anthers long-exserted at anthesis; filaments ? ciliate at base, the outer ones subulate, the inner ones with the anther-bearing cusp half as long as the expanded basal part and slightly shorter than the lateral cusps; lateral cusps much-exserted from the perianth; anthers yellow. Style well-exserted. Capsule unknown.

FLOWERING TIME. May–June
ECOLOGY. Near cultivation in desert.
DISTRIBUTION. Syria, known only from Qaryatein between Damascus and Palmyra.
NOTES. A little-known species in need of further study in the field. The type material was in a juvenile state and it is very likely that the disturbance caused by collecting resulted in the many sterile flowers which are a feature of the inflorescence. In many of its characters it resembles *A.ampeloprasum* but in the absence of any new material it is impossible to decide upon its relationships.

22. Allium oltense *Grossh.*, Fl. Kavk. ed.1,1:203(1928); Kollmann in Davis, Fl. of Turkey 8:167(1984). Type: Turkey, Erzurum prov., Oltu, *Grossheim* (lectotype TBI!).

ILLUSTRATIONS. Fl. Kavk. ed.2,2:t.13,f.4(1940). Plate 5D.

Bulb ovoid, 0.5–1 cm diam.; outer tunics membranous or coriaceous, sometimes breaking into parallel fibres; bulblets few, flattened, brown or purplish. Stem 19–40 cm. Leaves usually 2–3 (sometimes up to 4 in cultivated specimens), sheathing to about the lower third of the stem, linear, canaliculate, non-fistulose, 2–3 mm wide, margins, keel, and the prominent veins beneath, densely pilose-scabrid. Sheaths scabrid on the veins. Spathe 1-valved, ovate, abruptly narrowed to a beak 1.2–1.8 cm long, caducous. Umbel spherical, 1.5–2.5 cm diam., dense. Pedicels 0.7–1.2 cm long, slender; bracteoles present, numerous, white-scarious, up to 0.6 cm long. Perianth campanulate, umbilicate; segments green with white-transparent margins and some-times also white in the basal third, 2–3 mm long, smooth or very slightly pustulose on the keel, shiny, obtuse, the outer oblong-elliptic, the inner broader, elliptic. Stamens exserted; filaments ciliate at base, the outer ones simple, narrowly triangular, the inner ones with the anther-bearing cusp about half as long as the undivided basal part and slightly shorter than the lateral cusps; lateral cusps well-exserted from the perianth; anthers yellowish. Style exserted at anthesis, about 2 mm long. Capsule subglobose, about 2.5 mm long; seeds black, about 2–2.5 mm long. 2n=16.

FLOWERING TIME. July
ECOLOGY. Dry stony or rocky slopes, 1350–2200 m.
DISTRIBUTION. NE Turkey (recorded in Ağri, Artvin, Erzincan, Erzurum, Kars and Tunceli provinces).
NOTES. Although the type material is poor, and Grossheim's original description lacks fine detail, recently collected material, including some from the type locality (*A.Baytop et al.* ISTE 41196), has made it possible to describe *A.oltense* more fully. It is related to, and similar in appearance to, the next species, *A.rollovii*, differing mainly in flower size, leaf width and degree of scabridity of the leaves.

23. Allium rollovii *Grossh.*, Fl. Kavk. ed.1,1:205(1928), ed.2,2:119(1940); Kollmann in Davis, Fl. of Turkey 8:176(1984). Type: Turkey, Çoruh(Artvin) province, Artvin, *Grossheim*(?TBI or ERE).

Bulb oblong-ovoid; outer tunics greyish. Stem 20–40 cm. Leaves 1–2 mm wide, finely and densely scabrid–denticulate on the margin. Spathe caducous. Umbel c. 1 cm diam., hemispherical, few-flowered. Pedicels 0.5–0.7 cm long. Perianth segments white, 3–3.5 mm long, ovate-elliptic, smooth but with minutely ciliate-toothed mar-gins, keeled, the keel extended at the base into a tubercle. Stamens ?exserted or included; filaments ciliate at base, the inner ones with the anther-bearing cusp two thirds as long as the lateral cusps. 2n=16.

DISTRIBUTION. Turkey(recorded in Çoruh [Artvin] province).
NOTES. *A.rollovii* is a little-known species, perhaps collected only twice in the vicinity of Artvin. It appears to be fairly distinct in the 'flat-leaved' group in having very small umbels of white flowers, but nothing is known of its range of variation and little can be said of its affinities until it can be studied in greater detail. *N.Özhatay* ISTE 54708 (ISTE) probably represents this species, and the chromosome count given above is based on this collection. *A.rollovii* appears to be very closely allied to the next species, *A.talyschense*, and to *A.gramineum*; field studies are required in order to resolve the question of the relationships between these, and *A.oltense*.

24. Allium talyschense *Miscz. ex Grossh.*, Fl. Kavk. 1:204(1928); ed.2,2:121(1940); Vved. in Fl. URSS 4:251(1935); Engl. ed. 4:194(1968); Wendelbo in Rechinger, Fl. Iranica 76:56(1971). Type: Transcaucasia, Lenkoran, 1800–2400 m, *Misczenko* (not traced).

ILLUSTRATIONS. Fl. Kavk. ed.2,2:t.14,f.5(1940); Fl. Iranica 76:t.6,f.79(1971).

Bulb ovoid, 0.75–1.5 cm diam.; outer tunics pale-brown, coriaceous, sometimes breaking into fibres; bulblets few, small, yellow, shiny. Stem about 15–30 cm, smooth, prominently ribbed. Leaves 3–4, shorter than the inflorescence, sheathing the lower third of the stem, linear, canaliculate, non-fistulose, about 3 mm wide, scabrid on the margin and on the veins; sheaths smooth. Spathe 1-valved, broadly ovate with a beak about 1.3 cm long, caducous. Umbel usually spherical, 1.5–2.5 cm diam., dense. Pedicels 0.7–1 cm long, smooth; bracteoles present. Perianth campanulate; segments white with a dull purple median vein, about 5 mm long, oblong-lanceolate, acute, the outer keeled, scabrid on the outside and margins, the inner smooth. Stamens with the anthers included or about equalling the segments; filaments slightly ciliate at base, the outer ones triangular, the inner ones with the anther-bearing cusp about a third to a quarter as long as the ovate-triangular basal part and half as long as the lateral cusps; lateral cusps exserted from the perianth; anther colour unknown. Style included. Capsule subglobose, about 4 mm long.

FLOWERING TIME. June
ECOLOGY. Dry rocky places.
DISTRIBUTION. Transcaucasia, Talysh Mts.
NOTES. A poorly-known species in need of further study in the field. Morphologically it is close to *A.rotundum*. However, material from the type location, Lenkoran (LE!), suggests that it does differ from *A.rotundum* in having yellow-brown bulblets, leaves densely scabrid on the margins and sometimes also on the veins, and white flowers with purple mid-veins to the segments which are very scabrid on the keel and margins.

25. Allium gramineum *C.Koch* in Linnaea 22:239(1849); Grossh., Fl. Kavk. ed. 2,2:120(1940); Kollmann in Davis, Fl. of Turkey 8:175(1984). Type: Armenia, 'am Flusse Kharssakh im russischen Armenien auf Basaltboden', *C.Koch*(holotype B, ?destroyed).

A.fominianum Miscz. in Grossh. & Schischkin, Sched. Herb. Pl. Or. Exsicc. fasc.1-8:22,no.80(1924); Vved. in Fl. URSS 4:244(1935), Engl. ed.4:189(1968). Type: 'Transcaucasia, Tiflis, in lapidosis', 22 May 1919, *Grossheim*(isotypes K!, LE!). *A.rotundum* subsp. *ampeloprasoides* Miscz. in sched.
A.ampeloprasoides Miscz. ex Grossh., in Grossh. & Schischkin, op. cit. fasc.1-8:10,no.31(1924). Type: 'Transcaucasia pr. et dist. Tiflis, prope Kodzhory, 4000ft, in rupestribus', 29 June 1923, *Grossheim*(isotypes K!,LE).
A.fomini Miscz. ex Grossh., Fl. Kavk. ed.1,1:203(1928). Type: as for A.fominianum.
A.gramineum var. *ampeloprasoides* (Miscz. ex Grossh.)Grossh., Fl. Kavk. ed.2,2:121(1940).

Bulb ovoid, 0.8–1.2 cm diam.; outer tunics brown, coriaceous; bulblets, if present, yellowish-brown, helmet-shaped. Stem 20–40 cm, smooth(said to be glaucous). Leaves 2–3, shorter than the inflorescence, narrowly linear, flat, non-fistulose, 2–3 mm wide, minutely scabrid on the margin, sheathing the lower quarter to one third of the stem, apparentlyusually still green at flowering time; sheaths smooth, conspicuously dark-veined on a pale ground. Spathe caducous. Umbel spherical, (1.5–)2–3(–3.5)cm diam., dense. Pedicels 0.7–1.4 cm long, slightly unequal, smooth; bracteoles present. Perianth with a sweet-musty scent, shortly campanulate; segments white or greenish with a pinkish or greenish median stripe or vein, sometimes becoming purplish-tinged with age, 3.5–4 mm long, the outer ones oblong, subacute or obtuse, scabrid on the margins and keel, sometimes finely papillose all over the outer surface, inner ones broader, strongly keeled but smooth. Stamens included, or with slightly exserted anthers; filaments ciliate at base, the outer ones simple, linear-subulate, the inner ones with the anther-bearing cusp about a quarter to one third as long as the undivided basal part and half to two thirds as long as the lateral cusps; lateral cusps exserted from the perianth; anthers yellow or purplish before dehiscence. Style equal to or slightly exserted from the perianth. Capsule subglobose, 4 mm long. 2n=16.

FLOWERING TIME. May–June.

ECOLOGY. Stony soils and rocky places, 300–1300 m.

DISTRIBUTION. Transcaucasia; NE Turkey(recorded in Çoruh (Artvin) province.

NOTES. This is apparently allied to *A.rollovii* and *A.talyschense* and the three taxa require further investigation in the field in order to clarify their relationships.

26. Allium trachycoleum *Wendelbo* in Bot. Not. 122:35(1969); Wendelbo in Rechinger, Fl. Iranica 76:56(1971); Kollmann & Shmida in Israel Journ. Bot. 26,3:133(1977); Wendelbo in Townsend & Guest, Fl.of Iraq 8:156(1985); Kollmann in Davis, Fl. of Turkey 8:171(1984). Type: Iraq, Amadiya District, between Dohuk and Amadiya, above Sersang, 1200 m, 10–12 July 1957, *Rechinger* 11901 (holotype W, isotype K!).

ILLUSTRATIONS. Bot. Not. 122:30,f.1h(1969); Fl. Iranica 76:t.5,f.78(1971); Israel Journ. Bot. 26,3:134,f.2(1977).

Bulb ovoid, about 1.5 cm diam.; outer tunics blackish-brown, subcoriaceous, sometimes breaking into fibres; bulblets navicular, pointed at both ends, yellowish-brown or dark brown, shiny. Stem about (50–)60–100 cm, smooth. Leaves 5, shorter than the inflorescence, sheathing about the lower third of the stem, linear, canaliculate, non-fistulose, about 4 mm wide, scabrid on the margin and on the veins beneath; sheaths scabrid. Spathe caducous, 1-valved. Umbel spherical, 3–5(–6) cm diam., dense. Pedicels unequal, 1.5–3 cm long, smooth; bracteoles present, conspicuous, up to 1 cm long, silvery-membranous, laciniate. Perianth urceolate; segments white or greenish-white, 4–6 mm long, the outer elliptic-ovate, obtuse and often apiculate, scabrid-papillose on the outside and on the margins, the inner ovate, obtuse or subtruncate, smooth. Stamens with the anthers exserted at anthesis; filaments slightly ciliate at base, the outer ones triangular-subulate, the inner ones with the anther-bearing cusp half as long as the rectangular basal part and varying from about as long as to much shorter than the lateral cusps; lateral cusps about equalling or much-exserted from the perianth; anthers brown. Style well-exserted. Capsule subglobose, about 3 mm long. 2n=48[?].

FLOWERING TIME. June–July

ECOLOGY. Dry rocky places in oak forest, open stony slopes, 1050–2500 m.

DISTRIBUTION. N.Iraq (Amadiya District); S. & SE Turkey (Adana, Adiyaman, Bingöl, Hakkâri, Siirt & Tunceli provinces); W.Iran; Lebanon, Israel & Jordan (i.e. Mt. Hermon & Golan).

NOTES. The material from Turkey has smaller flowers, with perianth segments 4–5 mm long, than the Iraqi specimens in which they are 6 mm long. The material from Mt Hermon and the Golan area is noted by Kollmann & Shmida as having less scabrid perianth segments and inner filaments with the median cusp much shorter than the lateral cusps; in the type material the median cusp is about as long as the laterals. This taxon requires further study to determine if more than one entity is involved. N. & E. Özhatay ISTE 49447, 49448 & 54663 from Turkey: Gümüshane/Erzurum province, on the Kop Dağ Pass, may also belong to this species; however, the median cusp of the inner filaments is only about a third as long as the basal lamina. The flower colour of these specimens is not known.

27. Allium esfandiarii *Matine* in Iran. Journ. Bot. 4,2:166(1989). Type: Iran, Azarbayejan, Kalibar, Veinagh, Amirkhhan (forest), 1500 m, 5 July 1978, *Termeh, Moussavi & Habibi* s.n. (holotype EVIN).

ILLUSTRATIONS. Iran. Journ. Bot. 4,2:167,f.1(1989).

Bulb globose-ovoid, 1–2 cm diam.; outer tunics not known, but probably membranous; bulblets present, yellowish-brown, boat-shaped, apiculate. Stem up to 80 cm. Leaves 5, sheathing the lower third of the stem, linear, canaliculate, non-fistulose, about 3 mm wide, scabrid-papillose on the margins; sheaths scabrid-ciliate. Spathe 1-valved, with a broad base narrowed to a beak up to about 1 cm long, caducous. Umbel spherical or fasciculate-hemispherical, about 3 cm diam., dense. Pedicels unequal, about 1.5 cm long; bracteoles present, about 5 mm long, laciniate. Perianth urceolate-campanulate; segments purple with a darker mid-vein, 3–4 mm long, the outer narrowly elliptic, obtuse, cucullate, scabrid-papillose on the outside, the inner ovate-elliptic, retuse, nearly smooth. Stamens with exserted anthers; filaments papillose-ciliate at the base, the outer ones simple, triangular-subulate, the inner ones with the antherbearing cusp about a third as long as the elliptic-oblong undivided basal part and about half as long as the lateral cusps; lateral cusps well-exserted from the perianth; anthers dark purple before dehiscence. Style exserted. Capsule 3–4 mm long.

FLOWERING TIME. July
ECOLOGY. 'Forest', 1550 m.
DISTRIBUTION. Iran, known only from the type locality in Azarbayejan.
NOTES. This is described as being 'apparently most closely related to *A.qaradaghense* Feinbr. and *A.trachycoleum* Wendelbo, but differs markedly from these two species by the scabrid-ciliate leaf-sheaths and leaf margins, the colour of the tepals, which also have a strong nerve. From *A.qaradaghense* it furthermore differs in having the purplish-violet anthers, and from *A.trachycoleum* in the length of the inner filaments, which in *A.trachycoleum* have the lateral cusps the same length as the antheriferous cusp, whereas they are twice as long in *A.esfandiarii*'.

28. Allium pustulosum *Boiss. & Hausskn.* in Boiss., Fl. Or. 5:243(1882); Kollmann in Davis, Fl. of Turkey 8:170(1984). Type: Turkey, Malatya, Bey Dağ near Malatya, 17 Sept. 1865, *Haussknecht* s.n.(holotype G!, isotype JE!).

ILLUSTRATIONS. Fl. of Turkey 8:f.6,no.10(1984).

Bulb ovoid, 0.8–1.5 cm diam.; outer tunics greyish, membranous-subcoriaceous, splitting into rather weakly reticulated fibres, produced into a neck up to 4 cm long; bulblets flattened-ovoid, acuminate, pale brown, sometimes absent. Stem 20–60(–70) cm, smooth or slightly ribbed. Leaves 3–5, shorter than the inflorescences, linear, flat, non-fistulose, 2–3 mm wide, strongly scabrid on the margin and veins, sheathing the lower third to half of the stem; sheaths scabrid, prominently ribbed. Spathe deciduous. Umbel more or less spherical, densely-flowered, 1.5–3.5(–4) cm diam. Pedicels 0.5–2 cm long, smooth; bracteoles present, conspicuous, laciniate, silvery-membranous, up to 4 mm long. Perianth campanulate; segments white with a green median band, 2–3 mm long, narrowly ovate or elliptic-ovate, the outer acute, apiculate, slightly pustulose to scabrid-papillose, the inner obtuse, shorter and broader than the outer, smooth. Stamens with exserted anthers; filaments very sparsely and minutely ciliate-papillose at the base, the outer ones simple, triangular-subulate, the inner ones with the anther-bearing cusp shorter than the undivided basal part and the lateral cusps, sometimes only slightly so; lateral cusps well-exserted from the perianth; anthers yellow. Style 3 mm long. Capsule subglobose, 2–3 mm long. 2n=16.

FLOWERING TIME. June–August(–September)
ECOLOGY. Rocky(calcareous) mountain slopes, 1200–1900 m.
DISTRIBUTION. Turkey(recorded in the provinces of Elazığ, Malatya, Maraş, Mardin, Nevşehir, Sivas); ? N.Iraq.
NOTES. Possibly conspecific with *A.qaradagense*: see comments under the latter species.

29. Allium qaradagense *Feinbr.* in Pal. Journ. Bot., Jer. Ser. 3:8(1943); Wendelbo in Rechinger, Fl. Iranica 76:55(1971); Wendelbo in Townsend & Guest, Fl of Iraq 8:155(1985). Type: Iraq, Qara Dag, near Kani Takht, 1440 m, 12 Sept. 1933, *Eig & Feinbrun* s.n.(holotype HUJ!).

ILLUSTRATIONS. Pal. Journ. Bot., Jer. Ser. 3:f.2,15; Fl. Iranica 76:t.5,f.77(1971).

Bulb ovoid, c.1 cm diam.; outer tunics greyish, splitting into rather weakly reticulated fibres, produced into a long neck; inner tunics more conspicuously reticulate; bulblets rather small, brown, sometimes absent. Stem 30–60 cm, smooth. Leaves 3–4, shorter than the inflorescences, linear, flat, non-fistulose, 2–3 mm wide, scabrid on the margin, sheathing the lower third to two fifths of the stem; sheaths often papillose. Spathe deciduous. Umbel hemispherical to spherical, densely-flowered, 2–3 cm diam. Pedicels 0.8–1.5 cm long, smooth, bracteoles present, conspicuous, laciniate, 0.6–1 cm long, silvery-membranous. Perianth urceolate-campanulate; segments white, 3–3.5 mm long, subequal, elliptic-lanceolate or elliptic-ovate, the outer acute, apiculate, scabrid on the keel and margins, the inner obtuse, smooth. Stamens exserted; filaments minutely ciliate at the base, the outer ones simple, linear-subulate, the inner ones with the anther-bearing cusp half as long as the undivided basal part and half as long as the lateral cusps; lateral cusps well-exserted from the perianth; anthers yellow. Style ? not exserted. Capsule subglobose, 3–3.5 mm long.

FLOWERING TIME. June–July
ECOLOGY. Rocky or stony mountain slopes, in open oak forest, 800–1450 m.
DISTRIBUTION. Iraq, ? NW Iran.
NOTES. It is noted by Kollmann in Davis, Fl. of Turkey 8:170(1988) that this may be conspecific with *A.pustulosum* from E. & C.Anatolia. It was differentiated from this species by Feinbrun (l.c.) by its taller stems, larger umbel, longer pedicels and long lacerate bracts, but Kollmann notes that these characters are variable in *A.pustulosum*.

30. Allium willeanum *Holmboe*, Veg. Cypr. 45(1914); Lindberg f., Iter Cypr. 10(1946); Osorio-Tafall & Seraphim, List Vasc. Pl. Cypr. 22(1973); Meikle, Fl. Cypr. 2:1625(1985). Type: Cyprus, 'in the valley beyond Kaminaria', 1905, *Holmboe* 1070 (holotype G).

ILLUSTRATIONS. Holmboe, Veg. Cypr. f.9(1914).

Bulb ovoid, 1.5–3 cm diam.; outer tunics brown, membranous, splitting into parallel fibres; bulblets present, pale brownish-straw coloured, ovate, acuminate, produced beneath the tunics of the parent bulb or well above the bulb within the leaf sheaths. Stem 40–70(–100) cm. Leaves withered away by flowering time, 4–6, sheathing the lower quarter to third of the stem, linear, canaliculate, non-fistulose, keeled, 3–5 mm wide, scabrid-papillose on the margins; sheaths smooth. Spathe

1–valved, with a broadly ovate base abruptly narrowed to a slender beak up to 6 cm long, caducous. Umbel spherical, 2–5 cm diam., dense. Pedicels sub-equal, 1–2.3 cm long, glabrous; bracteoles present, minute, silvery-white. Perianth campanulate; segments whitish with a green mid-vein, 2–3 mm long, narrowly oblong, obtuse, truncate or emarginate, often slightly cucullate, papillose on the keel. Stamens with the anthers well-exserted; filaments papilloseciliate at the base, the outer ones simple, triangular-subulate, the inner ones with the anther-bearing cusp two thirds as long as to slightly shorter than the oblong undivided basal part and two thirds as long as to slightly shorter than the contorted lateral cusps; lateral cusps well-exserted from the perianth; anthers yellow before dehiscence. Style exserted. Capsule subglobose, 3 mm long; seeds black, about 2.5 mm long.

FLOWERING TIME. June–July

ECOLOGY. Dry rocky hillsides, crevices, dry river beds, roadsides, in waste ground and edges of fields, on limestone or igneous formations, sea level to 945 m.

DISTRIBUTION. Cyprus (recorded in most parts of the island).

NOTES. This is superficially very similar to *A.guttatum* but has linear non-fistulose leaves. It is more closely related to *A.cappadocicum* but the latter is a less robust plant with generally fewer, narrower, leaves and larger flowers which are usually pinkish with purple anthers; also, the bracteoles are fairly large and conspicuous unlike those of *A.willeanum* which are minute.

31. Allium alibile *A.Rich.*, Tent. Fl. Abyss. 2:330(1848); Baker in Fl. Trop. Africa 7:516(1898). Type: Ethiopia, Tigray Prov., Shire, near Tembella, *Quartin Dillon* (holotype P).

ILLUSTRATIONS. Plate 6A

Bulb globose, 1.5–2.5 cm diam.; outer tunics membranous, becoming fibrous with age, greyish; bulblets yellowish-brown. Stem (12–)30–65 cm, glabrous. Leaves 2–4, shorter than the inflorescence, sheathing approx. the lower third of the stem, linear, shallowly canaliculate, non-fistulose, keeled, (2–)3–5 mm wide, minutely scabrid on the margins; sheaths glabrous. Spathe 1-valved, with a broadly ovate base, abruptly contracted into a narrow beak exceeding the umbel. Umbel hemispherical, (1.5–)2.5–3 cm diam., dense. Pedicels up to 1.5 cm long, glabrous; bracteoles present, silvery-white, laciniate. Perianth campanulate-urceolate; segments white to pale pinkish-purple with a darker purple or green median vein, (3.5–)4 mm long, elliptic-ovate, the outer subacute or obtuse, apiculate, keeled, slightly scabrid on the outside, at least along the keel, the inner obtuse, glabrous. Stamens with anthers included or slightly exserted; filaments ciliate at the base, the outer simple, narrowly triangular, the inner ones with the anther-bearing cusp one third as long as the very widely expanded undivided basal part and one third to one half as long as the lateral cusps; lateral cusps about equal to or slightly exceeding the segments; anthers yellow or pinkish-purple. Style included. Capsule ovoid-subglobose, c. 2.5–3 mm long.

FLOWERING TIME. March–May (in Saudi Arabia), September–October (in Ethiopia).

ECOLOGY. On rocky grassy slopes, often among junipers, 1650–2134 m.

DISTRIBUTION. W Saudi Arabia; Ethiopia.

NOTES. The original description, based on Ethiopian material, does not contain all the relevant details but a good specimen, apparently representing the same taxon, was collected in 1975 [*Gilbert & Thulin* 998 (K!,UPS)] in the Blue Nile Gorge. This has made it possible to describe the species more fully and equate it with several specimens collected on the escarpment of western Saudi Arabia by *I.S.Collenette* and by

A.Arayed (K!). It is recorded as also occurring in Sudan, Jebel Marra, (see Norman in Journ. Bot. 62:135(1924) but the specimen (*Lynes* 37e,[BM!]) appears not to represent the same taxon. Further collections from Africa, including living material, of both *A.alibile* and the as yet unidentified species from Jebel Marra, are desirable.

32. Allium pervestitum *Klokov* in Kotov & Barbarich, Fl. RSS Ucr. 3:110,406(1950). Type: Ukraine, Zaporozhiensis District, near Mordvinovka by the Molocznaja river, 19 May 1932, *H.Bilyk* (holotype KW!)

ILLUSTRATIONS. Fl. RSS Ucr. 3:111,f.11(1950).

Bulb ovoid, 0.8–1.8 cm diam.; outer tunics brown, membranous or coriaceous, sometimes becoming fibrous, the fibres parallel or very weakly reticulate; bulblets dark purple or purplish-brown, plano-convex, ovate, 6–8 mm long. Stem 35–55 cm. Leaves 3–5, shorter than the inflorescence, sheathing the lower two thirds of the stem, linear, flat, non-fistulose, 2.5–6 mm wide, scabrid or minutely ciliate on the margin; sheaths conspicuously longitudinally green-veined. Spathe caducous. Umbel fastigiate to nearly spherical, 1.5–3.5 cm diam. Pedicels 0.5–1.5 cm long, smooth, yellow (? only on drying); bracteoles present, whitish-scarious, conspicuous. Perianth ovoid-campanulate; segments yellowish- white, 3–4 mm long; outer ovate-lanceolate with a scabrid-serrulate keel and scabrid margins, subacute; inner oblong, smooth, truncate or obtuse. Stamens slightly exserted at anthesis; filaments minutely ciliate at the base, the outer ones simple, narrowly lanceolate, the inner ones with the anther-bearing cusp a third to a sixth as long as the undivided basal part and a third to a quarter as long as the lateral cusps; anthers yellow. Style about 2 mm long. Capsule broadly ovoid, 3 mm long; seeds 2.5–2.75 mm long.

FLOWERING TIME. May
ECOLOGY. Saline soil near river
DISTRIBUTION. S.Ukraine
NOTES. Related to *A.albiflorum* but having smaller perianth segments, yellowish-white flowers and exserted anthers.

33. Allium albiflorum *Omelczuk* in Ukr. Bot. Zhur. 19,2:19(1962). Type: Crimea, Mt. Aj-Petri 10 June 1959, *N.Glagoleva* (holotype KW!)

ILLUSTRATIONS. Ukr. Bot. Zhur. 19,2:19,f.1(1968).

Bulb ovoid or nearly globose, 1–1.8 cm diam.; outer tunics coriaceous, sometimes breaking into parallel or weakly reticulate fibres; bulblets dark reddish-brown. Stem 35–70 cm. Leaves 3–4, shorter than the inflorescence, sheathing the lower third to half of the stem, linear, flat, canaliculate, non-fistulose, 2.5–4 mm wide, scabrid on the margin. Spathe shortly beaked, caducous. Umbel fastigiate-hemispherical, many-flowered. Pedicels 5–10(–12) mm long, ascending; bracteoles present, conspicuous, scarious, up to 5 mm long. Perianth ovoid-campanulate; segments white with a green median vein, becoming faintly purplish-pink with age, the outer 4–4.5 mm long, lanceolate, acute or subacute, conspicuously scabrid-crested on the keel, the inner subacute or obtuse, slightly broader. Stamens with anthers included; filaments slightly ciliate in the lower part, the outer ones simple, triangular-subulate, the inner ones with the anther-bearing cusp a quarter to a third as long as the undivided basal part and half as long as the lateral cusps; anthers yellow. Style ca. 1 mm long. Capsule sub-globose, about 3 mm long; seeds oblong, 2–2.5 mm long.

FLOWERING TIME. April–June
ECOLOGY. Stony mountain steppe, 600–900 m.
DISTRIBUTION. Crimea, N.Caucasus.
NOTES. Related to *A.pervestitum* but with larger perianth segments (3–4mm long in *A.pervestitum*), included stamens (exserted in *A.pervestitum*) and white flowers (yellowish in *A.pervestitum*).

34. Allium notabile *Feinbr.* in Pal. Journ. Bot., Jer. Ser. 3:7(1943); Blakelock in Kew Bull. 8:211(1953); Rawi in Dep. Agr. Iraq Tech. Bull. 14:186(1964); Wendelbo in Rechinger, Fl. Iranica 76:54(1971). Type: Iraq, Zawita, *Guest* 4960 (isotype K!).

ILLUSTRATIONS. Pal. Journ. Bot., Jer. Ser. 3:9,f.1,1a,1b(1943); Fl. Iranica 76:t.5,f.75(1971).

Bulb ovoid, 1.5–2.5 cm diam.; outer tunics reddish-brown, coriaceous, eventually splitting into parallel fibres; bulblets few, reddish-brown. Stem 35–95 cm, smooth. Leaves 3–5, sheathing the lower third of the stem, linear, canaliculate, non-fistulose, 2–4 mm wide, papillose on the margin. Spathe caducous. Umbel hemispherical, 2–4 cm diam., dense. Pedicels unequal, up to 2 cm long, very densely and conspicuously papillose-scabrid; bracteoles present, subulate, membranous, white. Perianth globose-campanulate, umbilicate at base; segments white, 2.5–3.5 mm long, elliptic-ovate, the outer cucullate, mucronate, keeled, scabrid-papillose, especially on the keel, the inner obtuse, almost smooth. Stamens with the anthers strongly exserted; filaments smooth, the outer ones triangular-subulate, entire, tridentate or tricuspidate, the inner ones with the anther-bearing cusp sub-equal to the undivided basal part and about two thirds as long as the lateral cusps; lateral cusps long-exserted from the perianth; anthers yellow. Style long-exserted. Capsule about 3 mm long, subglobose; seeds about 2.5 mm long.

FLOWERING TIME. June–July
ECOLOGY. Rocky mountainsides in pine and oak forest, 900–1100 m.
DISTRIBUTION. N.& E.Iraq, in Amadiya and Sulaimaniya Districts.
NOTES. A notable feature of *A.notabile* is that the pedicels are densely scabrid-papillose and it was this character, together with the pustulose perianth segments and the sometimes tricuspidate outer filaments, which suggested to N.Feinbrun that this species was particularly noteworthy.

35. Allium subnotabile *Wendelbo* in Bot. Not. 122:34(1969); Wendelbo in Rechinger, Fl. Iranica 76:55(1971). Type: Iran, Luristan, 30 km W. of Khorramabad, c.1280 m, 11 July 1966, *Archibald* 2671 (holotype GB!).

ILLUSTRATIONS. Bot. Not. 122:30,f.1g(1969); Fl. Iranica 76:t.5,f.76(1971).

Bulb ovoid, about 1 cm diam.; outer tunics brown, subcoriaceous, inner tunics reticulate-nerved and eventually becoming fibrous; bulblets purplish. Stem 37 cm, smooth. Leaves 4, shorter than the inflorescence, sheathing about the lower third of the stem, linear, canaliculate, non-fistulose, about 3 mm wide, papillose-scabrid on the margin and on the prominent veins beneath. Spathe caducous. Umbel subspherical, 2–2.5 cm diam., dense. Pedicels unequal, up to 8 mm long, slightly papillose-scabrid just below the flowers; bracteoles present, short, broad and lobed at the apex. Perianth campanulate-urceolate; segments white with an indistinct reddish median vein, subacute, cucullate, unequal, the outer shorter than the inner, about 5 mm long, elliptic-ovate, cymbiform, coarsely papillose-verrucose on the keel, the inner about 6 mm long, ovate, minutely scabrid-papillose on the outside. Stamens with the anthers included; filaments ciliate at base, the outer ones triangular-subulate, the inner ones with the anther-bearing cusp a third or less as long as the ovate-lanceolate undivided basal part and about two thirds as long as the lateral cusps; lateral cusps about equalling the perianth; anther colour unknown. Capsule about 3.5 mm long, subglobose.

FLOWERING TIME. June–July
ECOLOGY. On heavy clay over limestone formations, among *Quercus brantii*, 1280 m.
DISTRIBUTION. Iran, Luristan province, known only from the type locality.
NOTES. When describing the new species in 1969, Wendelbo compared it with *A.notabile* from which it is distinguished by the larger flowers, only slightly verrucose pedicels and included stamens.

36. Allium wendelboi *Matine* in Iran. Journ. Bot. 4,2:168(1989). Type: Iran, Zanjan to Gheydar, 35 km south of Soltaniyeh, Kuhhaye Gheydar, 2200–2650 m, 3 July 1974, *Termeh & Moussavi* s.n. (holotype EVIN).

ILLUSTRATIONS. Iran. Journ. Bot. 4,2:169,f.2(1989).

Bulb globose, 1–2 cm diam.; outer tunics brownish, coriaceous; bulblets present, few, yellowish-brown. Stem about 26 cm. Leaves shorter than the infloresence, 2–3, sheathing the lower third of the stem, linear, canaliculate, non-fistulose, about 5 mm wide; sheaths smooth. Spathe caducous, details unknown. Umbel spherical, 2.5–3 cm diam., dense. Pedicels unequal, about 1 cm long, densely pustulose; bracteoles present. Perianth urceolate-campanulate; segments whitish with a green mid-vein, about 4 mm long, the outer elliptic-lanceolate, subacute, cucullate, verrucose on the outside, especially on the keel and margins, the inner elliptic-ovate, slightly shorter, obtuse, less densely verrucose. Stamens with included or partially exserted anthers; filaments ciliate, the outer ones simple, triangular-subulate, the inner ones with the anther-bearing cusp about a quarter as long as the elliptic undivided basal part and about half as long as the lateral cusps; lateral cusps well-exserted from the perianth; anther colour unknown. Style exserted. Capsule about 4 mm long.

FLOWERING TIME. July
ECOLOGY. Habitat not noted, 2200–2650 m.
DISTRIBUTION. Iran, known only from the type locality.
NOTES. This is described as being related to *A.notabile* and *A.subnotabile*. From the former it is said to be 'distinguished by having the tepals with a strong green nerve, the included ciliate filaments, the inner filaments larger [broader-BM] than the tepals'. From *A.subnotabile*, the author notes that 'it differs clearly in having pustulose pedicels, the tepals with a strong green nerve, the outer tepals somewhat longer, not distinctly shorter, than the inner ones, as well as the inner filaments larger than the inner tepals. It is apparently very similar to these two species and it is impossible to discuss the relationship in any detail until further collections are available.

37. Allium drusorum *Feinbr.* in Pal. Journ. Bot., Jer. Ser. 3:8(1943); Mouterde, Nouv. Fl. Lib. et Syr. 1:265(1966). Type: Syria, 'Jebel Drouze, Tell Chihanne, (near Chaabo)', 21 June 1932, *Eig & Zohary* (holotype HUJ!).

ILLUSTRATIONS. Pal. Journ. Bot., Jer. Ser. 3:9,f.3(1943).

Bulb ovoid, 1.5 cm diam.; outer tunics brown, membranous, eventually splitting into parallel fibres, produced into a neck up to 3 cm long; bulblets unknown. Stem 25 cm. Leaves 2, sheathing the lower third of the stem, linear, canaliculate, non-fistulose, 2 mm wide, smooth. Spathe scarious, caducous. Umbel ovoid, somewhat fastigiate, 2–2.5 cm long, 1.5–2 cm wide, dense. Pedicels unequal, up to 1.5 cm long, the inner ones erect, the outer ones recurved; bracteoles present, membranous, white. Perianth ovoid, umbilicate at base; segments white with a straw coloured mid-vein, 3.5 mm long, oblong-lanceolate, obtuse, keeled, smooth. Stamens with the anthers exserted; filaments ciliate at the base, the outer ones simple, triangular-subulate, the inner ones with the anther-bearing cusp shorter than the undivided basal part and slightly shorter than the lateral cusps; lateral cusps exserted from the perianth; anther colour unknown. Style shortly exserted. Capsule details not recorded.

FLOWERING TIME. June
ECOLOGY. Unknown
DISTRIBUTION. Syria, known only from Jebel Druze.
NOTES. Apparently a rather short plant with only two narrow leaves and a white perianth with smooth segments; however, little is known of its variability and further material is required.

38. Allium ponticum *Miscz. ex Grossh.*, Fl. Kavk. ed.1,1:206(1928); Vved. in Fl. URSS 4:244(1935), Engl. ed. 4:189(1968); Kollmann in Davis, Fl. of Turkey 8:176(1984). Type: Transcaucasia, Tiflis, 17 June 1913, *Kozlowsky* 949 (lectotype TBI!) [inscribed by Misczenko 'A.ponticum m.'].

? *A.gracilescens* Somm. & Lev. in Acta Hort. Petrop. 13:51(1893). Type: Adzharia, Keda, 21 June 1890, *Sommier & Levier*(FI photo!). [See note below].

ILLUSTRATIONS. Fl. Kavk. ed.2,2:t.14,f.3,3a(1940).

Bulb ovoid, 1–1.5 cm diam.; outer tunics brown, membranous or coriaceous; bulblets purple, sometimes absent. Stem (30–)70–100 cm, smooth. Leaves 3–4, shorter than the inflorescence, narrowly linear, flat, canaliculate, non-fistulose, 2–7 mm wide, smooth or slightly scabrid on the margin, sheathing the lower quarter to one third of the stem; sheaths smooth. Spathe caducous. Umbel usually spherical, rarely hemispherical, 2.5–4.5 cm diam., many-flowered but fairly lax. Pedicels 0.8–2.2 cm long, slightly unequal, smooth; bracteoles present. Perianth shortly campanulate; segments pinkish-purple with a darker median stripe (or light rose, if *A.gracilescens* is included) with minute darker coloured (?glandular) pustules, 3–4 mm long, the outer ones oblong, obtuse, strongly scabrid on the outside, especially on the keel, inner ones broadly ovate, less strongly scabrid. Stamens included or shortly exserted; filaments ciliate at base, the outer ones simple, linear-subulate, the inner ones with the anther-bearing cusp about a third as long as the undivided basal part and half as long as the lateral cusps; lateral cusps exserted from the perianth; anthers purple before dehiscence. Style included or slightly exserted. Capsule globose, 4 mm long. 2n=16.

FLOWERING TIME. (May–)June–July.
ECOLOGY. Damp subalpine meadows, moist shady places, hillsides, and dry slopes; only altitude recorded is about 760 m.
DISTRIBUTION. Turkey(recorded in Rize and Artvin(Çoruh) provinces); Caucasus & Transcaucasus.
NOTES. It is not certain whether *A.gracilescens* Somm. & Lev. should be included

here as a synonym or retained as a distinct species. It appears that the only difference from *A.ponticum* is in the colour of the perianth, 'light rose' according to the authors; however the flower colour on the type specimen is now impossible to ascertain and it seems best, until further field studies have been undertaken, to mention *A.gracilescens* here under the umbrella of the better-known *A.ponticum*, although if they are subsequently shown to be conspecific the former epithet will take priority.

A.ponticum is clearly very closely allied to *A.rotundum* and further studies are desirable in order to ascertain whether the two can be satisfactorily distinguished.

39. Allium pseudocalyptratum *Mouterde* in Bull. Soc. Bot. France 100:347(1953); Mouterde, Nouv. Fl. Lib. et Syr. 1:264(1966); Kollmann & Shmida in Israel Journ. Bot. 26,3:131(1977); Kollmann in Rotem 15:61(1985). Type: Lebanon, 'sur les gres(?)?au-dessus de Zahle, près d'Ain Sawair[Saoua'ir]', 1800m, 17 July 1943, *Mouterde* (holotype P!).

ILLUSTRATIONS. Israel Journ. Bot. 26:132,f.1(1977); Rotem 15:62 & 63(1985).

Bulb ovoid, c. 1.5 cm diam.; outer tunics pale brown, membranous, extended at the apex into a long papery neck; bulblets dark purple, some carried adjacent to the parent bulb, others on long stipes and situated beneath the tunics high up on the neck of the bulb. Stem 50–85 cm, smooth. Leaves 3–4, shorter than the inflorescence, linear, flat, canaliculate, non-fistulose, c. 4–5 mm wide, scabrid-denticulate on the margin and sometimes also on the veins, sheathing the lower quarter to one third of the stem; sheaths smooth. Spathe not seen, presumably early-caducous. Umbel spherical, 2–4 cm diam., many-flowered, dense. Pedicels 0.5–1.5 cm long; bracteoles present, white, dissected. Perianth campanulate; segments white, with a green mid-vein, 4.5–5 mm long, scabrid on the outside, the outer ones lanceolate, obtuse, inner ones ovate-elliptic, obtuse. Stamens well-exserted; filaments glabrous, the outer ones simple, narrowly triangular, the inner ones with the anther-bearing cusp about half as long as the undivided basal part and about a third as long as the lateral cusps; lateral cusps long-exserted from the perianth; anthers yellow (brown according to Mouterde but probably based on dry material only). Style exserted. Capsule and seeds not seen.

FLOWERING TIME. June–July.

ECOLOGY. Rocky habitats, mainly on limestone formations, in open sites and in Cedar forest and sparse *Juniperus-Quercus* scrub, 1650–2800 m.

DISTRIBUTION. Lebanon, on the Lebanon range; Mt Hermon; it is also recorded by Kollmann & Shmida (op. cit. 1977) as occurring in N.Iraq; this observation was based on *Rawi* 23593(K!), a specimen previously identified by Wendelbo as *A.trachycoleum*. The specimen is not in good condition and is difficult to identify with any certainty but the possession of brown bulblets indicates that it is not *A.pseudocalyptratum*, in which they are dark purple.

NOTES. According to Kollmann and Shmida (op. cit. 1977) this is a variable plant, displaying a 'cline pattern in height of scapes (from 100 to 40 cm) and diameter of umbel(from 3–4 cm to 1.5 cm), both decreasing with rising altitude. Similar decrease in dimensions of scapes and umbels also occurs in rather dry habitats of the steppe country'.

The predominantly Turkish *A.calyptratum* also occurs on Mt Hermon but is easily distinguished from *A.pseudocalyptratum* on account of the fact that its anthers are not exserted.

40. Allium scorodoprasum *L.*, Sp. Pl. 1:297(1753); Reichenb., Ic. Fl. Germ. 10:23(1848); Boiss., Fl. Or. 5:232(1882); Aschers. & Graebn., Syn. Mitteleur. Fl. 3:100(1905), excl. var. *babingtonii*; Hegi, Ill. Fl. Mittel-Eur. 2:216(1909); Fiori, Nuova Fl. Anal. Italia 1:266(1923); Vved. in Fl. URSS 4:242(1935), Engl. ed. 4:187(1968); Stearn in Biol. Journ. Linn. Soc. 5:9(1973); Stearn in Fl. Europaea 5:65(1980); Kollmann in Davis, Fl. of Turkey 8:172(1984). Type: Sweden, Oland (lectotype Herb. Linn. 419/12, left specimen!).

A.*scorodoprasum* var. *viviparum* Regel, Monogr. Allium 43(1875). Type: no specimen cited.

A.*scorodoprasum* var. *ananthum* Beck, Fl. Nieder Österr. 1:166(1890). Type: no specimen cited.

ILLUSTRATIONS. Reichenb., Ic. Fl. Germ. 10:t.490,f.1073(1848); Hegi, Ill. Fl. Mittel-Eur. 2:f.334(1909); Butcher, New Ill. Brit. Fl. 649,f.(1961). Plate 6B.

Bulb ovoid, 1–2 cm diam.; outer tunics brown or greyish, membranous or subcoriaceous; bulblets present, oblong-ovoid, blackish-purple, produced beneath the bulb tunics and leaf sheaths. Stem 40–80 cm. Leaves 3–5, shorter than the inflorescence, sheathing approximately the lower third of the stem, linear, canaliculate, non-fistulose, keeled, 4–10(–20) mm wide, scabrid on the margin and midrib; sheaths glabrous. Spathe 1-valved, with an ovate base, narrowed to a beak which is up to twice as long as the base, caducous. Umbel lax, consisting of sessile dark violet bulbils and rather few (up to 20) flowers on long pedicels, usually somewhat irregular in shape, 1.5–3.5 cm diam. Pedicels unequal, up to 3(–4) cm long, smooth, often violet; bracteoles present, silvery-white. Perianth campanulate-ovoid; segments purple with a darker purple median vein, acute, 4–6 mm long, the outer ones oblong, keeled, scabrid along the keel, the inner oblong-ovate, slightly broader than the outer. Stamens with anthers included; filaments ciliate at the base, the outer simple, triangular-subulate, the inner ones with the anther-bearing cusp a third as long as the expanded undivided basal part and half as long as the lateral cusps; lateral cusps included or equalling the segments; anthers purplish. Style included. Capsule not seen, plant possibly sterile. 2n=16,32.

FLOWERING TIME. May–July

ECOLOGY. Fields, hedgerows, grassy places, scrub, vineyards, and waste places, usually at low altitudes.

DISTRIBUTION. N, C & E Europe, south to Bulgaria; Crimea; Caucasus; European Turkey & isolated localities in E & C Anatolia. This was formerly cultivated for culinary purposes and almost certainly the present-day wide distribution is a result of this; the production of many bulbils in the inflorescence ensures its survival long after organised cultivation has ceased.

NOTES. Closely related to *A.rotundum* and regarded by Stearn as a bulbil-forming derivative of this [Fl. Europaea 5:66(1981)], but see comments under *A.rotundum* [species no.41].

41. Allium rotundum *L.*, Sp. Pl. ed.2,1:423(1762); Reichenb., Ic. Fl. Germ. 10:24(1848); Regel, All. Monogr. 57(1875)p.p.; Boiss., Fl. Or. 5:233(1882); Fiori & Paol., Ic. Fl. Ital. 79(1898); Halàcsy, Consp. Fl. Graecae 3:247(1904); Coste, Fl. France 3:337(1906); Hayek, Prodr. Fl. Penins. Balcan. 3:43(1932); Vved. in Fl. URSS 4:246(1935), Engl. ed. 4:191(1968); Grossh., Fl. Kavk. ed.1,1:206(1928), ed.2,2:118(1940); Wendelbo in Rechinger, Fl. Iranica 76:54(1971).

ILLUSTRATIONS. Plates 6C–D; 7A.

Bulb ovoid, 1–2 cm diam.; outer tunics brown, coriaceous, splitting into fibres at the apex; bulblets present, very dark, usually blackish-purple, irregularly ovate, produced beneath the bulb tunics. Stem 20–70 cm. Leaves 3–5, shorter than the inflorescence, sheathing the lower third to half of the stem, linear, often canaliculate, non-fistulose, keeled, 2–7(–15) mm wide, scabrid on the margin; sheaths glabrous. Spathe 1-valved, about 1.5 cm long with an ovate base, narrowed to a short beak about equal in length to the base, caducous. Umbel spherical, 1–4(–5) cm diam., very dense. Pedicels smooth, unequal with the outer ones often very short and the inner ones much longer, up to 3 cm long, smooth; bracteoles present, numerous and conspicuous, silvery-white. Perianth campanulate–ovoid; segments pink to purple, sometimes with a darker purple median vein and the inner sometimes paler than the outer, the outer ones lanceolate or narrowly ovate, 4–6.5 mm long, keeled, scabrid along the keel, acute or subacute, the inner 4.5–7 mm long, elliptic or oblong-ovate, subacute or obtuse, often noticeably longer and broader than the outer. Stamens with anthers included or rarely slightly/partially exserted; filaments ciliate at the base, the outer simple, narrowly triangular, the inner ones with the anther-bearing cusp a quarter to a third as long as the expanded undivided basal part and one third to one half as long as the lateral cusps; lateral cusps slightly exceeding the segments; anthers usually yellow, rarely purple before dehiscence. Style included. Capsule ovoid-globose, about 4–5 mm long. 2n=16,32,38,40,48,64.

NOTES. *A.tmoleum* O.Schwarz in Fedde Repert. 36:71(1934) requires further investigation; it has acuminate perianth segments c.7 mm long, apparently rather pale in colour. It is based on *Schwarz* 933 (holotype B!) from Turkey, 'Lydia, Tmolus mons occidentalis, Ciplakdag supra Armutlu'. *A.cilicium* Boiss., Diagn. Ser.1,7:115(1846) is probably synonymous with *A.rotundum*; the type specimen was collected in southern Turkey, 'Monte Tauro Cilicicae', *Kotschy* 495 & 496 (G, K).

Stearn, in Ann. Mus. Goulandris 4:178(1978), logically treated *A.rotundum*, and the related *A.jajlae* and *A.waldsteinii* as subspecies of *A.scorodoprasum*. However, further research, at the Faculty of Pharmacy, Istanbul University, has shown that there are significant differences between *A.rotundum* and *A.scorodoprasum*, mainly concerning anatomical details of the bulb and its tunics, and of the leaves. On the basis of this work, Dr. N.Özhatay has expressed the opinion [pers. comm.] that these two taxa should be regarded as separate species, and for the purposes of this revision I have followed this course. However, it must be pointed out that the work involved only a limited area [European Turkey] within the whole, very large, range of distribution of these two taxa and it is clear that a much wider study is required to ascertain whether or not the criteria used to distinguish between them applies to all individuals throughout the range.

A.jajlae Vved. and *A.waldsteinii* G.Don have been distinguished from *A.rotundum* largely on account of their flower colour, uniformly pinkish in *A.jajlae*, uniform deep purple in *A.waldsteinii* and bicoloured in *A.rotundum*, with the outer segments deep purple and the inner whitish or very pale purple-pink. As already noted, Stearn treats these as subspecies of *A.scorodoprasum*, which is itself readily recognised on account of the presence of bulbils in the umbel. If *A.rotundum* and *A.scorodoprasum* are to be recognised as separate species, and if *A.jajlae* and *A.waldsteinii* are to be recognised at all, then it is considered more appropriate to regard these last two as subspecies of *A.rotundum* than of *A.scorodoprasum*. Unfortunately, in view of the fact that the differences between them are largely based on flower colour, it is impossible to make a thorough study of the group using herbarium material, which rarely bears field notes referring to flower colour. It is suggested that this whole *A.scorodoprasum-A.rotun-*

dum alliance, including *A.jajlae* and *A.waldsteinii*, would make an excellent topic for a postgraduate study.

For the purposes of this treatment, *A.jajlae* and *A.waldsteinii* are recognised as subspecies of *A.rotundum.*

Key to the three subspecies of *A.rotundum*:

1. Outer and inner perianth segments different in colour, the outer deep purple, the inner paler with white margins ....subsp. **rotundum**
   Outer and inner perianth segments the same colour 2
2. Perianth pink or pinkish-violet ....subsp. **jajlae**
   Perianth dark purple throughout ....subsp. **waldsteinii**

41a subsp. **rotundum** Type: S.Europe, 'Allium montanum capite rotundo', Haller (lectotype P! [Herb. *Haller* 40,no.7).

?*A.preslianum* Roem. & Schultes, Syst. 7:1132(1830). Type: Sicily, 'In arenosis maritimis ad Milazzo Siciliae' (not traced).
A.paterfamilias Boiss., Diagn. Pl. Or. Nov. ser.2,4:114(1859). Type: USSR, near R.Volga, *Becker* 160 (holotype G).
?*A.rotundum* var. *preslianum* (Roem. & Schultes)Regel, Monogr. All.:59(1875).
A.porphyroprasum Heldr. in Bull. Herb. Boiss. 6:394(1898). Syntypes: Greece, Attica, *Heldreich* s.n.[1851]; 800[1858]; 1387[1896] (isosyntype K!).
A.scorodoprasum subsp. *rotundum* (L.)Stearn in Ann. Mus. Goulandris 4:178(1978); Stearn in Fl. Europaea 5:65(1980); Kollmann in Davis, Fl. of Turkey 8:173(1984).

ILLUSTRATIONS. Reichenb., Ic. Fl. Germ. 10:t.492(1848); Hegi, Ill. Fl. Mittel-Eur. 2:f.335(1909); Fl. Iranica 76:t.5,f.74(1971); Bot. Not. 125:63,f.1A,2A,3F–G(1972). Perianth bicoloured, the outer segments deep purple, the inner much paler with white margins or almost wholly white with a pinkish-purple median vein.

FLOWERING TIME. May–July.

ECOLOGY. Dry sandy and rocky places, fields, vineyards, meadows and in scrub, usually on limestone formations or limestone-derived soils, sea level to 2250 m.

DISTRIBUTION. Widespread in S Europe, as far north as C Germany & Czech Rep.; Slovakia; European Russia; Crimea; Caucasus; Transcaucasus; Cyprus; Syria; Lebanon; Israel; W Iran; N Iraq; Turkey [widespread]; N Africa [recorded in Tunisia]. I have also seen living material from Morocco, Quezzane, *M.R.Salmon* s.n. which is apparently referable to this taxon although the bulblets are dark brown, not violet.

41b subsp. **jajlae** *(Vved.)B.Mathew*, comb. nov. Type: Crimea, Nikitskai yaila, 29 June 1924, *Wulff* (holotype LE).

A.jajlae Vved. in Bull. Univ. Asie Centr. 19:126(1934).
A.rotundum var. *melleum* Miscz. in Grossh., Fl. Kavk. ed.1,1:206(1928). Type: Armenia, Erevan, *Grossheim* (not traced).

Perianth uniformly pink or pinkish-purple.

FLOWERING TIME. May–July.

ECOLOGY. Fields, grassy places on sandy or clayey soils, sea level to 1400 m.

DISTRIBUTION. Crimea; Caucasus; N & E Turkey. In view of the difficulty in identifying dried material the exact distribution is not known.

41c subsp. **waldsteinii** *(G.Don)K.Richter*, Pl. Eur. 1:201(1890). Type: Turkey, Izmir province, 'in arvis et pratis Smyrnae', *G.Don* (holotype BM?).

A.waldsteinii G.Don, Monogr. All. 7(1827).
A.rotundum var. *waldsteinianum* Schultes & Schultes fil., Syst. Veg. 7:1011(1830).
A.rotundum var. *waldsteinii* (G.Don)Fiori in Fiori & Paol., Fl. Anal. Italia 1:196(1896).

ILLUSTRATIONS. Waldst. & Kit., Pl. Rar. Hung. 1:t82(1801).

Perianth uniformly deep purple.

FLOWERING TIME. June–July.
ECOLOGY. Fields, hillsides, grassy places, 1500–1900 m.
DISTRIBUTION. NE Italy; former Yugoslavia; NE Romania; N.Hungary; European Russia; Crimea; Caucasus; W , SW & E Turkey. The exact distribution is difficult to ascertain since the flower colour is not accurately retained in herbarium material and colour notes are usually lacking.

42. Allium asperiflorum *Miscz. ex Grossh.*, Fl. Kavk. ed.1,1:205(1928); Kollmann in Davis, Fl. of Turkey 8:174(1984). Type: Turkey, Çoruh (Artvin) province, Lomasen, *Grossheim* (not traced).

ILLUSTRATIONS. Fl. Kavk. ed.2,2:t.13,f.3(1940)(either a wrong caption or very inaccurate).

Bulb ovoid, 0.7–1 cm diam.; outer tunics brownish, membranous; bulblets dark purple. Stem 20–30(–40) cm. Leaves 1–2, narrowly linear, flat, non-fistulose, 1–3 mm wide. Spathe deciduous. Umbel spherical, densely-flowered, 2–3 cm diam. Pedicels 1–1.5 cm long, densely papillose; bracteoles present. Perianth ovoid-campanulate; segments purple, pale pink or white, 5–6 mm long, ovate, densely shaggy-scabrid or coarsely papillose on the exterior, the outer acute or subacute, the inner obtuse. Stamens included; filaments minutely ciliate at the base, the outer ones simple, the inner ones with the anther-bearing cusp shorter than the undivided basal part and the lateral cusps; anther colour, style and capsule details unknown. 2n=16,32.

FLOWERING TIME. June
ECOLOGY. Mountain meadows, 900–1450 m.
DISTRIBUTION. Turkey(recorded in the provinces of Adiyaman, Artvin, Elazığ, Erzincan and Tunceli).
NOTES. In the majority of its features *A.asperiflorum* appears to be somewhat similar to *A.rotundum*; the very shaggy-scabrid perianth segments enable it to be easily distinguished from the latter, but the reliability of this character must remain in question until a thorough field study can be undertaken. The material from around the type locality of Lomaşen in Artvin province of Turkey is fairly uniform in having densely scabrid segments [eg. *Sharman* 15013, 15014 (K!)], but some other specimens, [eg. ISTE 61641 (K!, ISTE!) from Tunceli province] have less-scabrid perianth segments and approach *A.rotundum* rather more closely. In view of the widely separated localities in which *A.asperiflorum* has been recorded the possibility must be considered that it merely represents unusual variants of *A.rotundum* in which the perianth segments are much more scabrid than normal.

43. Allium sintenisii *Freyn* in Ost. Bot. Zeitschr. 42:377(1892); Kollmann in Davis, Fl. of Turkey 8:190(1984). Type: Turkey, Erzincan Prov., Egin [Kemaliye], 'in agris otiosis ad Kota', 1 July 1890, *Sintenis* 2889 (holotype B!, isotype JE!,LD!).

ILLUSTRATIONS. Fl. of Turkey 8:183,f.7,no.14(1984).

Bulb ovoid, 0.7–1 cm diam.; outer tunics glossy, coriaceous; bulblets apparently absent. Stem 30–40 cm, smooth or verrucose towards the apex. Leaves 2–3, shorter than the inflorescence, flat, non-fistulose, 2–4 mm wide, margin, and sometimes the prominent veins, scabrid, sheathing the lower third to half of the stem; sheaths scabrid. Spathe caducous, not seen. Umbel spherical-ovoid, 2–3 cm diam., very dense. Pedicels (3–)5–10 mm long, smooth; bracteoles very conspicuous, white, membranous. Perianth tubular-campanulate or somewhat urceolate; segments pink with darker pink tips and median vein, 7–10 mm long, the outer ones lanceolate, tapering to an acute or subacute cucullate apex which is outward-curving (at least in the dried state), furnished with long (up to 1 mm) white papillae in the central part of the keel, the inner ones slightly longer, oblong, rounded at the apex, smooth. Stamens with the anthers included; filaments ciliate at base, the outer ones simple, narrowly triangular, the inner ones with the anther-bearing cusp about a quarter as long as the widely expanded undivided basal part and about two thirds as long as the lateral cusps; lateral cusps included well within the perianth; anthers yellow. Style included. Capsule about 4 mm long, seeds not seen. 2n=32.

FLOWERING TIME. June–July.
ECOLOGY. Seasonally moist mountain meadows, stony hillsides, about 1300 m.
DISTRIBUTION. E–C Turkey (recorded in Erzincan, Malatya and Sivas provinces).
NOTES. This is a rather distinctive species on account of the large pinkish flowers with acute outer perianth segments curved outwards at the tips and provided with very conspicuous long white papillae.

44. Allium erubescens *C.Koch* in Linnaea 22:242(1849); Vved. in Fl. URSS 4:246(1935), Engl. ed. 4:190(1968); Grossh., Fl. Kavk. ed.2,2:118(1940); Wendelbo in Rechinger, Fl. Iranica 76:58(1971); Kollmann in Davis, Fl. of Turkey 10:221(1988). Type: Transcaucasia, 'Am Strande des Kaspischen Meeres in der fruhern Herrschaft Kuba', *C.Koch* (not traced, probably destroyed).

A.brevicuspis Boiss., Diagn. Pl. Or. Nov. ser.2,4:114(1859); Boiss., Fl. Or. 5:241(1884). Type: Iran, 'in Persia boreali', *Buhse* (holotype G).
A.rudbaricum Boiss. & Buhse in Nouv. Mem. Soc. Nat. Mosc. 12:215(1860); Boiss., Fl. Or. 5:240(1884); Grossh., Fl. Kavk. ed.1,1:207(1928). Type: Iran, near Rudbar, May 1848, *Buhse* (lectotype G, isolectotype K!).
A.serrulatum forma *gracilior* Boiss. & Buhse in Nouv. Mem. Soc. Nat. Mosc. 12:214(1860). Type: Iran, 'Kaspische Kuste', 9 June 1848, *Buhse* (holotype G).

ILLUSTRATIONS. Fl. Iranica 76:t.5,f.73(1971). Plate 7B.

Bulb ovoid, 0.5–1.5 cm diam.; outer tunics brown or greyish-brown, coriaceous, splitting and becoming fibrous with age, produced into a distinct neck at the apex; bulblets present, dark purple, ovate in outline, acute, produced beneath the bulb tunics and on the neck beneath the leaf sheaths. Stem 20–40(–70) cm. Leaves 3–4, shorter than the inflorescence, sheathing the lower third to half of the stem, linear, shallowly canaliculate, non-fistulose, keeled, 2–7 mm wide, smooth or scabrid on the margin; sheaths glabrous. Spathe 1-valved, with a broadly ovate base, narrowed to a long beak about equal in length to the base, equal to or exceeding the umbel, caducous. Umbel spherical, 2–4 cm diam., dense. Pedicels up to 2.5(–3.5) cm long, smooth, purple; bracteoles present, conspicuous, silvery-white. Perianth campanulate; segments purplish-pink with a darker purple median vein, 5–7(–9) mm long, keeled, slightly scabrid along the keel, the outer narrowly lanceolate, with an outward-curving acute or acuminate apex, the inner oblong-lanceolate, acute. Stamens with anthers included; filaments ciliate at the base, the outer simple, triangular-subulate, the inner ones with the anther-bearing cusp one fifth to one third as long as the expanded undivided basal part and one third to one half as long as the lateral cusps; lateral cusps about equal to or slightly exceeding the segments; anthers yellow. Style included. Capsule ovoid-globose, about 4–5 mm long. 2n=24.

FLOWERING TIME. June–July

ECOLOGY. Dry sandy and rocky places, in meadows and in scrub, -20 to 1980 m.

DISTRIBUTION. E. Transcaucasus; Daghestan; Crimea; N.Iran; N.Turkey (Black Sea coast, near Samsun).

NOTES. Related to *A.rotundum* but fairly readily recognised by its bright pinkish-purple flowers with the acute segments curving outwards at the tips, carried on purple pedicels.

45. Allium calyptratum *Boiss.*, Diagn. Pl. Or. Nov. 1,13:30(1854); Boiss., Fl. Or. 5:242(1882); Feinbr. in Palestine Journ. Bot., Jer. Ser. 3:7(1943); Mouterde, Nouv. Fl. Lib. et Syr. 1:265(1966); Kollmann & Shmida in Israel Journ. Bot. 26,3:135(1977); Kollmann in Davis, Fl. of Turkey 8:175(1984); Kollmann in Rotem 15:69(1985). Type: Turkey, Hatay prov., 'in regione alpina montis Cassii(Akra Dağ) in consortio *A.cassii*', June 1846, *Boissier*(holotype G, isotype P!).

ILLUSTRATIONS. Israel Journ. Bot. 26:136,f.3(1977); Fl. of Turkey 8:165,f.6,no.15 (1984).

Bulb ovoid, 1–1.5 cm diam.; outer tunics dark brown, membranous (but inner tunics sometimes reticulate-fibrous), extended at the apex into a long fibrous neck; bulblets dark purple. Stem 30–60 cm. Leaves 3–4, shorter than the inflorescence, linear, flat, shallowly canaliculate, non-fistulose, 3–4.5 mm wide, sheathing the lower quarter to one third of the stem, smooth or slightly scabrid. Spathe suffused reddish-brown, membranous, ovate, narrowed to a long beak, caducous. Umbel hemispherical, 2–3 cm diam., rather lax. Pedicels 0.7–0.8 cm long; bracteoles present, conspicuous, dissected. Perianth broadly campanulate; segments shiny, white, with a green mid-vein, 4.5–5 mm long, the outer ones ovate-oblong, obtuse, scabrid on the keel, inner ones broadly ovate-ellipsoid. Stamens included or subexserted; filaments glabrous, the outer ones simple, narrowly triangular, the inner ones with the anther-bearing cusp about a third as long as the undivided basal part and a third to half as long as the lateral cusps; lateral cusps exserted from the perianth; anthers brown before dehiscence. Style included or slightly exserted. Capsule globose, 3 mm long. 2n=16,32.

FLOWERING TIME. May–July.

ECOLOGY. Montane regions, in rocky habitats, 1400–1700 m.

DISTRIBUTION. Turkey(recorded in Hatay(Antakya), Içel(Mersin), Konya and Maraş provinces); ? N.Syria; Mt. Hermon.

NOTES. This is recorded by Kollmann and Shmida(op. cit. 1977) as being extremely rare on Mt. Hermon. It appears to be allied to *A.rotundum* but can be readily distinguished by its white flowers and rather longer cusps to the inner filaments.

N.& E.Özhatay ISTE 48771 (ISTE!) from Turkey, Antalya province, between

Alanya and Anamur, is similar to *A.calyptratum* but differs in having the outer perianth segments densely papillose over the whole outer surface, and appears to have
brown rather than dark purple bulblets. This may represent a distinct and undescribed
taxon but requires further observation in the field.

46. Allium enginii *N.Özhatay & B.Mathew* in Kew Bull. 51(1): in press. Type:
Turkey, Içel, Mut to Kirobasi, 18 km from Mut, 15 June 1990, *N.& E.Özhatay* ISTE
61830 (ISTE!, K!).

ILLUSTRATIONS. Plates 7C–E.

Bulb globose, 0.7–1.5 cm diam.; outer tunics greyish, thinly membranous; inner
tunics white; bulblets absent. Stem 10–20 cm, somewhat flexuose, minutely papillose, ribbed, suffused dark purple in the lower part. Leaves 2(–3), usually exceeding
the inflorescence but sometimes equal in length, linear, non-fistulose, canaliculate
with the margins inrolled, 1.5–2.5 mm wide, prominently ribbed beneath, strongly
scabrid on the margin, sheathing the lower half to two thirds of the stem; sheaths
smooth, dark purple at the base. Spathe 1-valved, with a broadly ovate base, narrowed
to a slender beak c. 2.5 cm long, deciduous [spathe details taken from ISTE 46508
(ISTE!,K!)]. Umbel more or less conical, densely-flowered, 1–2 cm diam. Pedicels
0.5–0.8 cm long, smooth; bracteoles present, very conspicuous, broad and overlapping to form a silvery-membranous involucre around the base of the umbel, dentate
to laciniate. Perianth campanulate; segments reddish-purple with a darker median
line, finely papillose-scabrid over the whole abaxial surface, margins papillose, the
outer broadly ovate, 3.5–4.5 mm long, 3–4 mm wide, obtuse, the inner slightly longer
and narrower, 4–5 mm long, 2–2.5 mm wide, oblong, truncate. Stamens included; filaments minutely ciliate at the base, the outer ones simple, triangular, 1 mm wide at
the base, the inner ones 3-cuspidate with the anther-bearing cusp a quarter to a third
as long as the undivided basal lamina and a quarter to a third as long as the lateral
cusps, the basal lamina elliptical and about as wide as the adjacent inner perianth segment; lateral cusps equalling the perianth or very slightly exserted; anthers yellow.
Style shortly exserted at anthesis, c. 1–1.5 mm long. Capsule subglobose, 3 mm long;
seeds 2.5 mm long. 2n=16.

FLOWERING TIME. June
ECOLOGY. Dry rocky places, limestone formations, 850–1350 m.
DISTRIBUTION. Turkey(recorded in the province of Içel).
NOTES. A distinctive species which has the overall appearance of *A.reuterianum*
but has 'flat' solid leaves rather than fistulose ones and is in fact not closely related.
The rather spherical bulb with greyish membranous tunics is more characteristic of
the species in Sect. *Melanocrommyum* (eg. *A.orientale* Boiss.) than of Sect. *Allium*,
but the features of the stamens clearly identify the new species as a member of the latter section. Although it has several characters in common with the very widespread
and variable *A.rotundum*, it differs markedly in having a spherical bulb with greyish-
membranous tunics and no bulblets (reddish-purple bulblets are a notable feature of
A.rotundum), a flexuose stem which is overtopped, or at least equalled, by the leaves,
and a more or less conical umbel subtended by wide silvery bracteoles which give the
impression of an involucre. The perianth segments are densely papillose-scabrid all
over the outer surface whereas those of *A.rotundum* are usually scabrid only along the
keel.

Although it seems that the new species normally has two leaves, vigorous specimens show the unusual character of producing a third very narrow leaf within the

sheath of the upper leaf.

47. Allium dregeanum *Kunth*, Enum. Pl. 4:382(1843); Baker in Fl. Cap. 5:407(1897); Wilde-Duyfjes, Revis. Allium in Africa 75(1976); Type: South Africa, 'Port Natal et Afrique Meridionale', *Drège* Herb. Cap. 8660a (isotype G).

ILLUSTRATIONS. Wilde-Duyfjes, Revis. Allium in Africa 76,f.12(1976). Plate 8A.

Bulb ovoid, 1–2 cm diam.; outer tunics brown, membranous, sometimes splitting lengthways; bulblets elongate-ovoid, acuminate, produced beneath the tunics of the parent bulb, brown. Stem 30–90 cm, smooth but sometimes prominently ribbed. Leaves about 3–5(–6), shorter than the inflorescence, sheathing the lower third to half of the stem, linear, flat or canaliculate, non-fistulose, keeled, 2–6(–9) mm wide, glabrous, conspicuously and closely veined; sheaths glabrous. Spathe 1-valved, the valve ovate at the base, tapering abruptly to a beaked apex, exceeding the umbel, caducous. Umbel with flowers only or flowers and bulbils mixed, spherical or hemispherical, 3–5 cm diam., densely-flowered when wholly floriferous or rather lax when bulbils present; bulbils sessile, ovate or fusiform, about 5 mm diam., brown. Pedicels up to 2.5 cm long, smooth, rather thick; bracteoles present, conspicuous, up to about 7 mm long, white-membranous, somewhat laciniate. Perianth campanulate-urceolate; segments whitish or pale pink with a green or purple median vein, 5–8 mm long, ovate or ovate-oblong, the outer subacute, mucronate, keeled, sparsely scabrid on the keel, the inner obtuse or rounded-truncate, bluntly mucronate, smooth. Stamens with anthers included; filaments ciliate near the base, the outer ones simple, narrowly triangular, the inner ones with the anther-bearing cusp half as long as the widely expanded undivided basal part and about two thirds as long as the lateral cusps; lateral cusps included or very slightly exserted; anthers yellow. Style included. Capsule subglobose, about 3–4 mm long; seeds about 3 mm long, black.

FLOWERING TIME. October–December(i.e. spring–summer).
ECOLOGY. Sandy soils, dolerite hills, 100–1830 m.
DISTRIBUTION. South Africa, widely distributed although not common.
NOTES. Some doubt has been expressed as to whether *A.dregeanum* is a true native of South Africa or an old introduction which has become naturalised. The latter scenario seems unlikely since (1) it has proved impossible to equate the S.African material exactly with any of the other taxa in section *Allium* and (2), it is recorded that in 1652 (a very early date in terms of European settlement) the Hottentots were gathering the plant. William J. Burchell collected specimens of it in the early 19th century and made the observation [*Travels in the interior of Southern Africa* 1:296(1822)] that 'The green leaves of a kind of onion, growing here wild, were plucked by many of the Hottentots, and boiled with their meat.' Unfortunately the specimen to which this statement relates, *Burchell* 1547, Carnarvon Division, 'Schiet Fontein', 9th September 1811(K!), is immature; the inflorescences are malformed and the flower buds show doubling of all the floral parts, but it is possible to confirm that it belongs to section *Allium*, with tricuspidate inner filaments.

It is probable that there is an earlier epithet for *A.dregeanum*; several *Allium* species were described from South Africa by G.Don, Monograph of the genus Allium (1827) and at least one of these, *A.synnotii* G.Don, possesses all the same basic characteristics as *A.dregeanum*. However, no specimen has been traced ['This plant was brought home by Mr.Synnot, from the Cape of Good Hope, in 1825, and flowered in 1826'] and it is considered unwise and unhelpful to suggest a name change for *A.dregeanum* on the basis of an incomplete description alone; the same comments apply to *A.verrucosum* G.Don.

48. Allium baeticum *Boiss.*, Diagn. Pl. Or. Nov. 1,7:113(1846); Willk. & Lange, Prodr. Fl. Hisp. 1:210(1862); Regel, All. Monogr.:56(1875); Coutinho in Bol. Soc. Brot. 13:102(1896); Coutinho, Fl. Port.:129(1913), ed.2:153(1939); Maire, Fl. Afr. Nord 5:260(1958); Wilde-Duyfjes, Revis. Allium in Africa:78(1977); Stearn in Ann. Mus. Goulandris 4:177(1978); & in Fl. Europaea 5:65(1980). Types: Spain, "in regione calida provinciae Malacitanae"(Malaga), *Boissier*(?G); "Granatensis" (Granada), *Boissier* (?G); "circa Gades"(Cadiz), *Fauché*(?G).

A.durieanum Gay in Ann. Sci. Nat. Bot. 3,8:218(1847). Types: Algeria, near Bone, June 1844, *Durieu de Maisonneuve*(P); & Aug.1844, *Durieu de Maisonneuve*(lectotype P).

A.ampeloprasum L. var. *durieanum* (Gay)Bonnet & Barratte, Cat. Pl. Vasc. Tunis:413(1896).

A.baeticum var. *occidentale* Coutinho in Bol. Soc. Brot. 13:102(1896). Types: Portugal, Villa Franca, Monte Gordo, *Da Cunha* s.n.(?); near Lisbon, Serra de Monsanto, *Da Cunha* s.n.(?).

A.margaritaceum Smith var. *papillosum* Lindberg in Acta Soc. Sci. Fenn., N.S. B,1,2:32(1932). Type: Morocco, Great Atlas, near Asni, *Lindberg* 3622(holotype MPU).

A.savarini Sennen in Sennen et al., Cat. Fl. Rif Or., Melilla, Iberica:120(1933), nom. nud., based on *Sennen & Mauricio* s.n. from Morocco(BM!).

A.baeticum var. *laeve* Maire & Weiller in Maire, Fl. Afr. Nord 5:261(1958). Type: as for *A.baeticum*.

ILLUSTRATIONS. Maire, Fl. Afr. Nord. 5:f.893(1958); Wilde-Duyfjes, Revis. Allium in Africa: f.13(1977).

Bulb ovoid, 1–3 cm diam.; outer tunics deep golden brown, coriaceous at first, the outer ones breaking up and forming a thick mat of reticulate and parallel fibres, produced into a long neck 2.5–9 cm long; bulblets up to 5(–6), flattened-ovoid, sessile or stalked, shiny dark brown, sometimes absent. Stem 30–90 cm, longitudinally ribbed (at least when dry), glabrous. Leaves 3–6(–10), usually shorter than the inflorescence, sheathing the lower half of the stem, withering away at or before flowering time, linear, flat or shallowly canaliculate, non-fistulose, keeled, (1–)3–7 mm wide, glabrous. Spathe 1-valved, 2–6.5 cm long, membranous, ovate at the base, tapering to a long beak, caducous. Umbel usually spherical, 3–6 cm diam. Pedicels slender, 1–2.5 cm long, unequal, smooth or minutely scabrid; bracteoles present, membranous, 3–5 mm long, dissected at the apex. Perianth nearly cylindrical or urceolate; segments white or pale purplish-pink with a green mid-vein, 3–5 mm long; outer narrowly oblong, acute, scabrid-papillose, especially on the keel; inner narrowly oblong, obtuse or subacute, smooth. Stamens with included anthers, about equalling the perianth; filaments papillose-serrulate near the base, the outer ones simple, narrowly ovate-triangular at the base, subulate at the apex, the inner ones flattened and much wider at the base than the outer; anther-bearing cusp about half as long as the undivided basal part and half as long as the lateral cusps; lateral cusps slightly exserted from the perianth; anthers purplish-pink before dehiscence then creamy-yellow. Style 3–6 mm long. Capsule ovoid or nearly globose, about 4 mm long; seeds black, about 3 mm long. 2n=32.

FLOWERING TIME. May–August.

ECOLOGY. Rock crevices, fields, sandy places and in scrub of *Juniperus* and *Quercus*, sea level to 2500 m.

DISTRIBUTION. S.Spain, S.Portugal, Morocco, Algeria, Tunisia.

NOTES. Unfortunately it has not been possible to acquire living specimens of this species, and very little herbarium material has been seen. Although it is grouped here with the 'reticulate-tunic, flat-leaved' species, such as *A.dictyoscordum* and *A.longicollum*, it is possibly more closely allied to *A.ampeloprasum*; further investigations should be made concerning its relationships.

49. Allium longicollum *Wendelbo* in Bot. Not. 121:272(1968); Wendelbo in Rechinger, Fl. Iranica 76:49(1971). Type: Afghanistan, Kandahar Prov., 44 km NE of Qala Bist, *Rechinger* 34761(holotype W, isotype K!)

ILLUSTRATIONS. Bot. Not. 121:271,f.1,B–C(1968); Fl. Iranica 76:t.5,f.64(1971).

Bulb ovoid, 1.5–2 cm diam.; outer tunics brown, forming a thick mat of reticulate fibres, produced into a long neck; bulblets absent(?always). Stem 16–30 cm, ribbed, the ribs sparsely pubescent with retrorse hairs. Leaves 3, carried close together low down on the stem, withering away at or before flowering time, linear, somewhat flattened, non-fistulose, 1–3 mm wide, retrorse-pilose along the nerves. Spathe with 2–3 ovate, acuminate valves 5–7 mm long, papery with green veins. Umbel hemispherical to nearly spherical, 1–2 cm diam. Pedicels 0.5–1 cm long, smooth; bracteoles present, minute. Perianth campanulate; segments white with a green mid-vein, 3–4 mm long; outer lanceolate or oblong-lanceolate, obtuse or subacute, cucullate, smooth or slightly scabrid on the margin; inner elliptic-oblong or narrowly obovate, obtuse, cucullate, smooth. Stamens included, the tip of the anthers approximately equalling the perianth segments; filaments smooth, the outer ones simple, narrowly ovate-triangular at the base, subulate at the apex, the inner ones twice as wide at the base as the outer; anther-bearing cusp half as long as the undivided basal part and half as long as the lateral cusps; lateral cusps equalling or slightly exserted from the perianth; anthers violet before dehiscence. Style 4 mm long. Capsule subglobose, about 4 mm long.

FLOWERING TIME. April–May.
ECOLOGY. Desert, stony slopes, 900–1800m.
DISTRIBUTION. SW & SE Afghanistan.
NOTES. Related to *A.dictyoscordum* from Kopet Dag but with smaller perianth segments and retrorse-hairy stems and leaves.

50. Allium dictyoscordum *Vved.* in Not. Syst. Herb. Hort. Bot. Petrop. 5:90(1924); Wendelbo in Rechinger, Fl. Iranica 76:48(1971); Matin in Iranian Journ. Bot. 5(1):7–8(1991). Type: Turkmenistan, Kopet Dag, Germab, *Antonov* (holotype LE).

ILLUSTRATIONS. Fl. Iranica 76:t.5,f.63(1971); Iranian Journ Bot. 5(1):7(1991).

Bulb ovoid, 1.2–2.5 cm diam.; outer tunics brown, reticulate-fibrous, produced into a long neck; bulblets absent. Stem 50–70 cm. Leaves withering away at or before flowering time, 4–6, clustered at the base of the stem, linear, canaliculate, non-fistulose, about 3 mm wide, margin strongly scabrid. Spathe with 2 shortly beaked valves, shorter than the umbel, persistent. Umbel spherical or hemispherical, 1.5–2 cm diam., dense. Pedicels 0.7–1.5 cm long; bracteoles present, white. Perianth ovoid-campanulate; segments white with a green mid-vein, 5–6 mm long, the outer ovate, subacute, cucullate, scabrid on the keel, the inner oblong-obovate, obtuse, smooth. Stamens included, the tip of the anthers approximately equalling the perianth segments; filaments smooth, the outer ones simple, narrowly ovate-triangular at the base, subulate at the apex, the inner ones twice as wide at the base as the outer and wider than the perianth segments; anther-bearing cusp half as long as the undivided basal part and about half as long as the lateral cusps; lateral cusps slightly exserted from the perianth; anther colour unknown. Style 4 mm long. Capsule subglobose, about 5 mm long.

FLOWERING TIME. May–June.
ECOLOGY. Saline clay soils in *Artemisia* association, "lower mountain zone".
DISTRIBUTION. Turkmenistan, Kopet Dag range; Iran.
NOTES. Related to *A.longicollum* and sharing the same bulb tunic features but a much taller plant with less hairy leaves and longer perianth segments.

51. Allium gypsodictyum *Vved.* in Fl. Uzbek. 1:543 & 453(1941). Type: C.Asia, Pamir-Alai, 'in gypsaceis montium Babatag', S slope opposite Chagam, 10 July 1936, *Lepeshkin & Mukhamedzhznove* 496 (holotype TAK).

ILLUSTRATIONS. Fl. Uzbek. 1:t.66,f.3(1941).

Bulb ovoid, 1.5–2 cm diam.; outer tunics reddish-brown, reticulate-fibrous, produced into a distinct neck; bulblets few, brown, smooth. Stem 20–25 cm, smooth. Leaves 2–3, filiform, canaliculate above, non-fistulose, about 0.5–1 mm wide, scabrid, sheathing the lower half of the stem; sheaths scabrid. Spathe caducous. Umbel hemispherical, 3–4.5 cm diam., few-flowered, lax. Pedicels 1.5–2.5 cm long; bracteoles present. Perianth campanulate; segments pink with a darker purple median vein, about 5 mm long, acute, smooth, the outer lanceolate, keeled, the inner linear-lanceolate, slightly longer. Stamens with anthers included at anthesis; filaments glabrous, the outer ones simple, narrowly triangular, the inner ones with the anther-bearing cusp about a quarter the length of the basal lamina and about half as long as the lateral cusps; lateral cusps slightly exserted from the perianth; anther colour unknown. Style included. Capsule details unknown.

FLOWERING TIME. June–July.
ECOLOGY. Mountain slopes on gypsum formations.
DISTRIBUTION. C.Asia, Pamir-Alai Mts.
NOTES. Although one would expect this to be related to other similar central Asian species such as *A.filidens*, *A.crystallinum* and *A.margaritiferum*, this is described as having 'efistulose' leaves which would indicate a lack of close affinity with these species and point to a link with *A.dictyoscordum*. It is necessary that this character is carefully checked since it is very difficult to observe in dried material, and particularly so in this case since the leaf width is less than 1 mm. It is unfortunate that living material has been unavailable, since this would allow the anatomy of the leaf to be determined more accurately than from dried specimens.

52. Allium gomphrenoides *Boiss. & Heldr.* in Boiss., Diagn. Pl. Or. Nov. 1,7:114(1846); Regel, All. Monogr.:65(1875); Boiss., Fl. Or. 5:243(1882); Halàcsy, Consp. Fl. Graecae 3:248(1904); Hayek, Prodr. Fl. Penins. Balcan. 3:44(1932); Rech. fil., Fl. Aegaea: 716(1943); Greuter & Rech. fil. in Boissiera 13:159(1967); Stearn in Ann. Mus. Goulandris 4:187(1978); & in Fl. Europaea 5:68(1980). Syntypes: Greece, Peloponnese, Taygetos Mts., above Androuvista, *Heldreich* (syntype G, isosyntype K!); Peloponnese, *Bory* (syntype G).

ILLUSTRATIONS. Plate 8C.

Bulb ovoid, 0.6–1 cm diam.; outer tunics brown, reticulate-fibrous; bulblets usu-

ally absent. Stem 8–24 cm, smooth. Leaves 2–3, shorter than the inflorescence, narrowly linear, non-fistulose, about 1 mm wide, smooth, sheathing the lower sixth to
one third of the stem; sheaths smooth. Spathe 2-valved, 0.5–1 cm long, valves ovate-
lanceolate, beaked, persistent. Umbel hemispherical with the flowers fastigiate, very
dense, 1–2.5 cm diam. Pedicels 2–8 mm long, slightly unequal, smooth, without
bracteoles. Perianth campanulate; segments pink or purple, 5.5–7 mm long, elliptic-
oblong, usually obtuse, sometimes emarginate, smooth. Stamens included; outer filaments narrowly lanceolate, smooth, the inner ones very broad and minutely ciliate at
the base, the anther-bearing cusp about a third as long as the undivided basal part and
half as long as the lateral cusps; lateral cusps included within the perianth; anthers
yellow. Style included. Capsule subglobose, 3.5 mm long. 2n=16.

FLOWERING TIME. April–June.
ECOLOGY. Rocky places, low altitudes.
DISTRIBUTION. S Greece, southern Peloponnese and Kithira Is.
NOTES. A distinctive species with relatively large flowers in small very dense
umbels.

53. Allium rubrovittatum *Boiss. & Heldr.* in Boiss., Diagn. Pl. Or. Nov.
2,13:29(1853); Regel, All. Monogr.:68(1875); Boiss., Fl. Or. 5:234(1882); Halàcsy,
Consp. Fl. Graecae 3:247(1904); Hayek, Prodr. Fl. Penins. Balcan. 3:42(1932); Rech.
fil., Fl. Aegaea: 716(1943); Stearn in Ann. Mus. Goulandris 4:187(1978); & in Fl.
Europaea 5:68(1980). Type: Greece, Crete, Kordaliotikon, June 1846, *Heldreich*
(holotype G, isotype K!).

ILLUSTRATIONS. Plate 8B.

Bulb ovoid, 0.5–1 cm diam.; outer tunics brown or blackish, coriaceous; bulblets
usually absent. Stem 2–20 cm, smooth but slightly ribbed. Leaves 3–4, shorter than
the inflorescence, linear-filiform, fistulose, about 1–1.5 mm wide, smooth, sheathing
the lower one third of the stem; sheaths smooth. Spathe 2-valved, up to 0.5 cm long,
valves ovate, beaked, persistent. Umbel hemispherical with the flowers mostly fastigiate, dense but sometimes reduced to only 2 or 3 flowers, 0.5–2 cm diam. Pedicels
1–6(–12) mm long, unequal, smooth, with a few small silvery-membranous bracteoles. Perianth campanulate; segments reddish-purple with a white margin, 3.5–4 mm
long, narrowly ovate, obtuse or subacute, papillose. Stamens just included to partially exserted; outer filaments narrowly lanceolate, sparsely ciliate at the base or smooth,
the inner ones very broad and smooth, the anther-bearing cusp about a third as long
as the undivided basal part and half as long as the lateral cusps; lateral cusps slightly
exserted at anthesis; anthers purple before dehiscence. Style equalling the perianth or
slightly exserted. Capsule subglobose, 3 mm long; seeds 2 mm long. 2n=16.

FLOWERING TIME. May–June.
ECOLOGY. Dry rocky places on calcareous formations, sea level to 600 m.
DISTRIBUTION. Crete and Karpathos Is.
NOTES. This is endemic to Crete and Karpathos Is. and is related to *A.junceum* but
with much smaller flowers and a slightly different structure to the inner filaments.

54. Allium junceum *Smith* in Sibth. & Smith, Fl. Graecae Prodr. 1:226(1809) & Fl.
Graeca 4:19(1823); Regel, All. Monogr. 71(1875); Boiss., Fl. Or. 5:238(1882);
Holmboe, Veg. Cypr. 46(1914); Osorio-Tafall & Seraphim, List Vasc. Pl. Cypr.
22(1973); Kollmann in Davis, Fl. of Turkey 8:190(1984); Meikle, Fl. Cypr.

2:1620(1985).

Bulb ovoid, 0.5–1.5 cm diam.; outer tunics brown, membranous, splitting at the apex and base into parallel fibres; bulblets usually absent. Stem 8–35(–60) cm, glabrous, prominently veined. Leaves 2–4, shorter than, or sometimes equalling or just exceeding the inflorescence, filiform, fistulose, 0.5–2 mm wide, glabrous, sheathing the lower third to half of the stem; sheaths glabrous. Spathe broadly ovate, 1-valved or splitting irregularly into 2 ovate lobes, shortly acuminate, 1–1.5 cm long, persistent. Umbel spherical-fastigiate or ovoid-fastigiate, (0.5–)1–2(–3) cm diam., compact and dense but sometimes with as few as 2 flowers. Pedicels glabrous, up to 4 mm long; bracteoles present, linear-filiform, silvery, hidden by the persistent spathe. Perianth narrowly campanulate; segments pale purple with a dark purple median vein or deep purple with paler margins, 6–7 mm long, ovate, acute, subequal or the outer slightly shorter than the inner, the outer keeled, minutely papillose-scabrid on the keel, margins smooth or ciliate-dentate. Stamens with included anthers; filaments glabrous, the outer lanceolate-subulate, the inner ones with the anther-bearing cusp a sixth to a quarter as long as the expanded undivided basal part and about half as long as the lateral cusps; lateral cusps included, sometimes each with a sharp tooth on the outer edge giving a slightly 5-cuspidate appearance; anthers purple before dehiscence [?sometimes yellow — see note below re. *A.trachyanthum*]. Style included. Capsule ovoid or subglobose, 3–5 mm long; seeds angular, about 3 mm long, black. 2n=16.

NOTES. *A.trachyanthum* Griseb., Spic. 2:395(1844) may also belong here. It was described from Turkey, near Bolu and, from the description, appears to be very similar to *A.junceum* although the latter has not been recorded as far north in Turkey as Bolu province and is phytogeographically a Mediterranean species. The perianth segments of *A.trachyanthum* are said to be obtuse and the anthers yellow, these features being at variance with those of *A.junceum*. I have not traced the type material and its status must therefore remain in doubt for the present.

A.junceum is allied to *A.rubrovittatum* from Crete but has considerably larger flowers and the median cusp of the inner filaments is much shorter in relation to the basal lamina. In Fl. of Turkey 8:191(1984) Kollmann recognised two infraspecific taxa based on the relative lengths of the inner and outer perianth segments and the absence or presence of teeth on the lateral cusps of the inner filaments. The characters appear to hold good for the majority of specimens but material collected by M.Johnson in Antalya province, Turkey, and cultivated at Kew, suggests that these characters are not entirely reliable and thus require careful assessment over the whole range of the species. Furthermore, Meikle, Fl. of Cyprus 2:1621(1985), states that in Cypriot material 'the lateral cusps may vary from being quite entire to deeply laciniate-lobed'.

The two subspecies which have been recognised are as follows:

1. Outer perianth segments about as long as the inner; inner filaments with toothed lateral cusps ....subsp. **junceum**
   Outer perianth segments shorter than inner; inner filaments with ....
   entire lateral cusps ....subsp. **tridentatum**

54a subsp. **junceum**. Type: Cyprus, 'in insula Cypro', *Sibthorp* (holotype OXF).

A.chauveli Boissieu in Bull. Soc. Bot. France 43:290(1896). Type: Turkey, Antalya province, 'sables maritimes a Chiralu (Cirali) et rocailles en montant a la Chimere', *Boissieu* (holotype P).

A.ferrinii Pamp. in Bull. Soc. Bot. Ital. 1925:142(1925). Type: Rhodes, Mt. Psalido, 3–4 June 1924, *Ferrini* (not traced).

ILLUSTRATIONS. Sibth. & Sm., Fl. Graeca 4:t.322(1823); Fl. of Turkey 8:183,f.7,no.15(1984). Plate 9A.

Perianth segments subequal; each of the two lateral cusps of the inner filaments usually provided with a sharp tooth on the outer edge, the filaments thus appearing to be 5-cuspidate. 2n=16.

FLOWERING TIME. April–May.
ECOLOGY. Sandy and rocky places, cliffs, dry pastures, garrigue, often on limestone formations, sea level–1000 m.
DISTRIBUTION. Cyprus, Rhodes, SW Turkey (recorded in provinces of Antalya, Aydin, Muğla).

54b subsp. **tridentatum** *Kollmann, N.Özhatay & Koyuncu* in Notes R.B.G. Edinb. 41:248(1983); Kollmann in Davis, Fl. of Turkey 8:191(1984). Type: Turkey, Antalya province, Konyaalti, 10 m, 17 May 1936, *Tengwall* 669 (holotype K!).

ILLUSTRATIONS. Notes R.B.G. Edinb. 41:251,f.3B(1983); Fl. of Turkey 8:183,f.7,no.16(1984).

Outer perianth segments shorter than the inner; lateral cusps of the inner filaments not toothed, the filaments thus appearing 3-cuspidate. 2n=16.

FLOWERING TIME. May.
ECOLOGY. Rocky places near sea, sea level to 60 m.
DISTRIBUTION. S.Turkey, recorded in Antalya province.

55. Allium heldreichii *Boiss.*, Diagn. Pl. Or. Nov. 3,4:116(1859); Regel, All. Monogr.:64(1875); Boiss., Fl. Or. 5:237(1882); Halàcsy, Consp. Fl. Graecae 3:248(1904); Hayek, Prodr. Fl. Penins. Balcan. 3:44(1932); Stearn in Ann. Mus. Goulandris 4:188(1978); & in Fl. Europaea 5:68(1980). Type: Greece, Mt Olympus, *Heldreich* (holotype G).

ILLUSTRATIONS. Plate 9B.

Bulb ovoid-subglobose, 0.8–1 cm diam.; outer tunics blackish, membranous; bulblets usually absent. Stem 20–60 cm, smooth. Leaves 2–4, shorter than the inflorescence, terete, fistulose, 0.5–3 mm wide, smooth, sheathing the lower quarter to one third of the stem; sheaths smooth. Spathe with 2 valves 1–1.5 cm long, valves lanceolate, acuminate, persistent. Umbel hemispherical with the flowers fastigiate, very dense, 2.5–4.5 cm diam. Pedicels 0.5–1.5 cm long, unequal, smooth, without bracteoles. Perianth campanulate; segments pink with a green mid-vein, 8–11 mm long, lanceolate, acute, smooth. Stamens included; filaments minutely ciliate at the base, the outer ones simple, narrowly lanceolate, the inner ones with a very broad base, the anther-bearing cusp about a quarter as long as the undivided basal part and two to three times shorter than the lateral cusps; lateral cusps included within the perianth; anthers purple before dehiscence. Style included. Capsule subglobose, 4–5 mm long; seeds 2–3 mm long. 2n=14,16.

FLOWERING TIME. June–August.
ECOLOGY. Rocky mountain slopes, 700–2000 m.
DISTRIBUTION. N.Greece.

NOTES. Stearn, in Flora Europaea 5:68(1980) comments that this species resembles *A.schoenoprasum* in its outward appearance, but this is of course superficial and there is no close relationship between them; the tricuspidate inner filaments immediately identify it as a member of section *Allium*. The very large pink flowers carried fastigiately in dense umbels make it immediately recognisable.

56. Allium ilgazense *N.Özhatay* in Notes R.B.G. Edinb. 44:147(1986); Kollmann in Davis, Fl. of Turkey 10:222(1988). Type: Turkey, Kastamonu province, Ilgaz Dağ, SW of Geyik pass 29 July 1983, *N.& E.Özhatay* ISTE 51918(holotype ISTE!).

ILLUSTRATION. Notes R.B.G. Edinb. 44:147,f.2(1986).

Bulb ovoid, 0.5–1 cm diam.; outer tunics black, membranous; bulblets not known. Stem (25–)30–45(–50) cm, purplish at base. Leaves 2–4, shorter than the inflorescence, linear, semiterete, fistulose, 2–3 mm wide, glabrous, sheathing the lower third of the stem. Spathe persistent, 2–3-valved, the valves ovate, acute, c.10mm long, membranous. Umbel globose, densely-flowered, 2–3.5 cm diam. Pedicels subequal, 0.5–1.5 cm long, tuberculate in the upper part, bracteoles present, white, membranous, minute. Perianth campanulate, pink; segments 6.5–9.5 mm long, the outer ones slightly shorter than the inner, ovate-oblong, subacute, cymbiform, verrucose-scabrid on the keel, the inner ones lanceolate, truncate or emarginate. Stamens slightly exserted; outer filaments simple, triangular-subulate, ciliate at the base; inner ones with the anther-bearing cusp slightly shorter than the undivided basal part and the lateral cusps; lateral cusps exserted from the perianth; anthers purplish. Style exserted. Capsule ovoid, 4–5 mm long. 2n=16.

FLOWERING TIME. July–August
ECOLOGY. In coniferous woods and rocky places, sometimes on calcareous soils, 1450–2000 m.
DISTRIBUTION. N.Turkey (recorded only in the province of Kastamonu).
NOTES. In Flora of Turkey 10:222(1988), N.Özhatay suggested that this is "allied to *A.jubatum* Macbride from which it differs by the size, scabridity and colour of the perianth, and to *A.heldreichii* Boiss. (N.Greece), from which it is separated by its longer filaments and purplish anthers."

57. Allium jubatum *Macbride* in Contr. Gray Herb. n.s.56:7(1918); Stearn in Ann. Mus. Goulandris 4:187(1978); Stearn in Fl. Europaea 5:68(1980); Kollmann in Davis, Fl. of Turkey 8:191(1984). Syntypes: Bulgaria, near Kalofer, Sept. 1844, *Pichler* 226 (holotype G); Turkey, Bithynia, *Thirke* (holotype G, photo K!).

? *A.ciliatum* C.Koch in Linnaea 19:11(1846); non Cyr.
A.cristatum Boiss., Fl. Or. 5:237(1882); Hayek, Prodr. Fl. Penins. Balcan. 3:43(1932);
   non S.Watson(1879). Type: as for *A.jubatum*.

ILLUSTRATION. Fl. of Turkey 8:183,f.7(1984). Plates 9C; 10A.

Bulb ovoid, 0.8–1.5 cm diam.; outer tunics greyish or blackish, membranous; bulblets absent or few. Stem 15–40 cm. Leaves 2–3, shorter than the inflorescence, linear, semiterete, fistulose, ribbed, 1–3 mm wide, glabrous, sheathing the lower third to half of the stem; sheaths closely and prominently ribbed, smooth. Spathe 2-valved, the valves 1–1.2 cm long, ovate, shortly acuminate, membranous, persistent. Umbel globose or subglobose, densely-flowered, 1.5–2(–2.5)cm diam. Pedicels unequal or subequal, 2–3(–8) mm long, minutely scabrid-papillose; bracteoles present, narrow-

113

ly linear, white, membranous. Flowers smelling of dung; perianth oblong-campanulate; segments 6–7.5 mm long, the outer ones shorter than the inner, purplish, oblong-lanceolate, acute or subacute, cymbiform, scabrid on the keel, the inner ones paler with purple tips, narrowly oblong, somewhat expanded and more or less truncate at the apex and furnished with many small teeth, papillose outside. Stamens included; outer filaments simple, triangular-subulate, the inner ones ciliolate at the base, with the anther-bearing cusp a third to half as long as the undivided basal part and two thirds as long as the lateral cusps; lateral cusps slightly exserted from the perianth; anthers pale yellow. Style exserted. Capsule subglobose, about 4 mm long; seeds black, 3 mm long. 2n=16.

FLOWERING TIME. May–June.

ECOLOGY. In calcareous rocky places, 200–1000 m.

DISTRIBUTION. C. Bulgaria(? extinct), NW Asiatic Turkey(recorded in the provinces of Bolu, Bursa, Bilecik, and Çankiri).

NOTES. An easily identified species on account of the conspicuously toothed inner perianth segments. It is thought that it may be extinct in Bulgaria since it has not been found there since the original collection in 1844. However, in Turkey, although not very widespread, it is apparently not uncommon in some of the localities where it does occur; Dr Neriman Özhatay has informed me that it grows in very large numbers in the Yedigöller Natural Park and is therefore protected.

One of the living collections at Kew (*Koyuncu* 8573) has flowers which smell of dung, but it is not known how consistent this is within the species.

58. Allium aucheri *Boiss.*, Diagn. Pl. Or. Nov. 1,7:116(1846); Boiss., Fl. Or. 5:237(1882); Vved. in Fl. URSS 4:238(1935), Engl. ed. 4:184(1968); Wendelbo in Rechinger, Fl. Iranica 76:50(1971); Kollmann in Davis, Fl. of Turkey 8:189(1984). Types: Turkey, Armenia, *Aucher* 2192 (lectotype G, isolectotype K!, P!).

A.carduchorum C.Koch in Linnaea 22:237(1849). Type: Turkey, Erzurum, *Brant*(holotype B).

A.brevipes Ledeb. Fl. Ross. 4:165(1852). Type: Armenia, 'prope Erivan', *Steven*(holotype LE, photo K!).

A.caerulescens Boiss., Diagn. Pl. Or. Nov. 2,4:115(1859) & Fl. Or. 5:233(1882). Type: Turkey, Erzurum, *Calvert* (isotype CGE?), non *A.caerulescens* G.Don, Monogr. Allium 34(1827).

A.ledschanense Conrath & Freyn in Bull. Herb. Boiss. 4:190(1896). Type: Caucasus, 'Somchetia, in montis Ledschan prope Privoluge', c. 2000 m, *Conrath* 119 (holotype GZU!).

A.cyclospathum Freyn in Bull. Herb. Boiss. ser.2,1:288(1901). Type: Turkey, Van, 'in tractu Warack[Erek] Dagh', *A.Kronenburg* 69 (holotype BRNM or B?).

A.faniae Stearn in Ann. Mus. Goulandris 4:158(1978), new name for *A.caerulescens* Boiss.(1859).

ILLUSTRATIONS. Fl.URSS 4:t.14,f.3(1935); Fl. Iranica 76:t.5,f.66(1971); Fl. of Turkey 8:183,f.7,no.13(1984). Plate 10B.

Bulb ovoid, (0.5–)0.8–1 cm diam.; outer tunics greyish-brown, subcoriaceous, splitting lengthways into triangular strips; bulblets apparently absent. Stem 40–70 (–80) cm, smooth, rather wiry. Leaves 2–4, much shorter than the inflorescence, terete, only very slightly flattened on the adaxial surface, fistulose, 1.5–3(–4) mm wide, smooth, sheathing the lower half of the stem, the free part of the lamina short, less than 15 cm long; sheaths smooth and rather inflated at the apex, conspicuously ribbed. Spathe 2-valved, the valves ovate, abruptly narrowed to a short beak, membranous, purplish-tinged, persistent. Umbel spherical with the flowers sometimes rather fastigiate, 2–3 cm diam., very dense. Pedicels 3–6 mm long, smooth; bracteoles absent or very few, filiform, white, membranous. Perianth campanulate-conical at anthesis; segments violet blue with a green or violet median vein, 7–9 mm long, oblong-lanceolate, tapering to an acute or subacute cucullate apex, smooth or scabrid on the keel. Stamens with the anthers included; filaments ciliate at base, the outer ones simple, narrowly triangular, the inner ones with the anther-bearing cusp about a sixth to a quarter as long as the widely expanded undivided basal part and about a quarter to a half as long as the lateral cusps; lateral cusps included well within the perianth; anthers purple before dehiscence. Style included. Capsule ellipsoid, about 6 mm long; seeds 2.5 mm long. 2n=16.

FLOWERING TIME. June–July.
ECOLOGY. Seasonally moist mountain meadows, grassy hillsides, 1800–2550 m.
DISTRIBUTION. E.Turkey; NW Iran; Transcaucasus.
NOTES. This is a distinctive species with its violet-blue flower colour and short leaves which develop in spring, rather than in autumn as in many of the species. It resembles *A.hierochuntinum* which is, however, a plant of dry steppe with a more southern distribution in Israel, Jordan and Syria; the latter has reticulate-fibrous bulb tunics and shorter perianth segments, and the inner filaments have the central cusp about a third the length of the basal lamina (only one sixth to one quarter in *A.aucheri*).

59. Allium nevsehirense *Koyuncu & Kollmann* in Israel Journ. Bot. 27:90(1978); Kollmann in Davis, Fl. of Turkey 8:184(1984). Type: Turkey, Nevşehir prov., Göreme Valley, 17 August 1975, *Shmida* (holotype HUJ!).

ILLUSTRATIONS. Israel Journ. Bot. 27:90,f.1a–e(1978); Fl. of Turkey 8:183,f.7,no.6(1984). Plate 11A.

Bulb ovoid, 0.8–1.5 cm diam.; outer tunics greyish, membranous; bulblets, if present, almost as large as the parent bulb, elongate, acuminate, shortly stalked, greyish or yellowish. Stem 30–80 cm, smooth. Leaves 2–4, shorter than the inflorescence, subterete, fistulose, canaliculate, 2–5 mm wide, smooth or scabrid-serrate on the margin, sheathing the lower third of the stem; sheaths smooth. Spathe 2-valved, the valves 5–8 mm long, ovate, shortly acuminate, persistent. Umbel spherical, 1.5–3 cm diam., dense. Pedicels 5–13 mm long, smooth, markedly swollen at the base and apex; bracteoles present, white, membranous. Perianth oblong-campanulate or broadly ellipsoid; segments greenish-yellow or green, 3–4 mm long, smooth, the outer oblong-ovate or oblong-obovate, obtuse, cymbiform, the inner broadly elliptic-ovate, subacute or obtuse. Stamens exserted; filaments ciliate at the base, the outer ones simple, narrowly triangular, the inner ones with the anther-bearing cusp about a third to half as long as the expanded undivided basal part and about equal in length to or slightly longer than the lateral cusps; all cusps exserted from the perianth; anthers yellowish, about 1.5 mm long (?rarely purple before dehiscence: see *P.H.Davis* 68835 from Turkey, Sivas prov., E!). Style exserted by 1–1.5 mm. Capsule globose, about 3 mm long. 2n=16.

FLOWERING TIME. July–August.
ECOLOGY. Dryish rocky hillsides, vineyards, on volcanic formations, gypsum or shale, 800–900 m.
DISTRIBUTION. C.Turkey (recorded in Erzincan, Gümüşane, Kayseri, Nevşehir, and

Sivas provinces).

NOTES. The dense umbel of smallish green flowers makes this a fairly distinctive species; it is most closely related to *A.phanerantherum* but differs in having smaller flowers and smaller anthers, and an ovate-elliptic ovary (oblong in *A.phaneran-therum*). Furthermore, the central, anther-bearing, cusp of the inner filaments is equal to or longer than the lateral cusps whereas in *A.phanerantherum* it varies from half as long as to slightly shorter than the lateral cusps. It is also somewhat similar to *A.artvinense* but has smaller tighter umbels of flowers.

60. Allium phanerantherum *Boiss. & Hausskn.* in Boiss., Fl. Or. 5:235(1882); Feinbr. in Palestine Journ. Bot., Jer. Ser. 3:11(1943); Mouterde, Nouv. Fl. Lib. et Syr. 1:267(1966); Wendelbo in Rechinger, Fl. Iranica 76:48(1971); Kollmann in Notes R.B.G. Edinb. 33:305(1974); Kollmann & Shmida in Israel Journ. Bot. 26,3:135(1977); Kollmann in Davis, Fl. of Turkey 8:182(1984); Kollmann in Rotem 15:69(1985); Wendelbo in Townsend & Guest, Fl. of Iraq 8:152(1985); Kollmann in Feinbr., Fl. Palaestina 4:91(1986). Types: Turkey, Gaziantep prov., 'in dumetis montis Ssoffdagh Syriae borealis', 914 m, 27 June 1865, *Haussknecht*(syntype G, photo K!); Lebanon, 'in Libano prope Aleih', 762 m, *Schweinfurth*(syntype G, photo K!).

A.davisianum Feinbr. in Palestine Journ. Bot., Jer. Ser. 3:13(1943) emend. Kollmann in Notes R.B.G. Edinb. 31:121(1971). Type: Israel, Samaria, Wadi Beidan, near Nablus, 10 July 1942, *P.H.Davis* 4920 (holotype E).
A.descendens auct. plur., non L.(1753).

ILLUSTRATIONS. Palestine Journ. Bot., Jer. Ser. 3:9,f.4 & 19,f.23(1943)[as *A.davisianum*]; Mouterde, Nouv. Fl. Lib. et Syr. 1, Atlas: t.83,f.6(1966); Fl. Iranica 76:t.5,f.69(1971); Fl. of Turkey 8:183,f.7,no.3(1984); Rotem 15:68(1985). Plate 9D.

Bulb ovoid, 1–2 cm diam.; outer tunics greyish–brown or blackish, membranous, sometimes splitting into parallel fibres; bulblets few, elongate, produced on slender stalks above the parent bulb and beneath the tunics, yellowish or sometimes purplish. Stem (30–)50–100 cm, smooth or scabrid. Leaves 2–5, shorter than the inflorescence, linear, semiterete, fistulose, canaliculate, 1–5 mm wide, glabrous or scabrid-papillose on the veins, sheathing the lower quarter to third of the stem, almost entirely shriv-elled away by flowering time; sheaths smooth. Spathe usually splitting into several acuminate lobes 5–13 mm long, whitish-membranous, usually persistent at least until anthesis, sometimes with only one shortly-beaked lobe. Umbel spherical or rarely ellipsoid, (1.5–)2–4.5 cm diam., dense. Pedicels 5–20 mm long, minutely papillose at the apex; bracteoles absent or present. Perianth cylindrical or urceolate, umbilicate at the base; segments pale green or purple, sometimes white at the base, usually with a darker mid-vein, sometimes coloured darker towards the tips, 4–5 mm long, smooth, the outer oblong-elliptic, obtuse, cymbiform, the inner ovate-elliptic, subacute. Stamens exserted; filaments ciliate at the base, the outer ones simple, triangular or tri-angular-subulate, the inner ones with the anther-bearing cusp two thirds as long as to slightly shorter than the expanded undivided basal part and varying from about half as long as to only slightly shorter than the lateral cusps, all the cusps exserted from the perianth; anthers purple before dehiscence or yellow, about 2.2 mm long. Style included or exserted, purple or green. Capsule ellipsoid, 4 mm long; seeds 2.5 mm long.

FLOWERING TIME. June–August.

ECOLOGY. Dryish stony or grassy hillsides and in sparse oak forests, on limestone or sandstone formations, sometimes on clay or conglomerate soils, screes, 100–1885 m.

DISTRIBUTION. S.Turkey (recorded in Adana, Antalya, Gaziantep, Hatay(Antakya), İçel(Mersin), Isparta, Maraş and Muğla provinces); W.Iran; Iraq; W.Syria; Lebanon; Jordan; Israel; possibly also in Saudi Arabia.

NOTES. In Turkey *A.phanerantherum* is used on a local scale for flavouring food.

A.phanerantherum subsp. *deciduum* Kollmann & Koyuncu [Notes R.B.G. Edinb. 41:263(1983)] is in need of further study in the field. It was based on a specimen collected by *Eig & Zohary* in the Amanus Mts., Adana Prov., Turkey, on 1 July 1932 (holotype HUJ). The distinguishing characteristics of subsp. *deciduum* and subsp. *phanerantherum* are given in Flora of Turkey 8:182(1984) as follows:

Spathe persistent, split into several lobes;
    bracteoles absent ....subsp. **phanerantherum**
Spathe deciduous; bracteoles present ....subsp. **deciduum**

However, the Kew duplicate of one of the cited specimens [*P.H.Davis* 16343!], and *Eig & Zohary* s.n. 30 June 1932 (HUJ!), of subsp. *deciduum*, are in fact in possession of their spathes at anthesis. This matter therefore requires further investigation since the spathe characters are thought to have considerable significance in the delimitation of species and groups of species within sect.*Allium*. The absence or presence of bracteoles also needs detailed investigation in the field.

61. Allium damascenum *Feinbr.* in Palestine Journ. Bot., Jer. Ser. 3:16(1943); Mouterde, Nouv. Fl. Lib. et Syr. 1:266(1966). Type: Syria, near Damascus, 14 May 1931, *Zohary* (holotype HUJ!).

ILLUSTRATIONS. Pal. Journ. Bot., Jer. Ser. 3:15,f.5;19,f.26(1943).

Bulb ovoid, 1.5 cm diam.; outer tunics greyish, membranous; bulblets few, ovoid, tapering to an acuminate apex. Stem 37–50 cm. Leaves 2–3, green at anthesis, shorter than the infloresence, sheathing the lower third of the stem, linear, semiterete, canaliculate in the lower part, fistulose, 2–3 mm wide, with prominent ribs, especially the two lateral ones which are nearly wing-like, scabrid on the two lateral ribs. Spathe 2-valved, the valves ovate at the base, acuminate or shortly beaked, 1.2–1.5 cm long, scarious, persistent. Umbel globose, 4–5 cm diam., rather lax. Pedicels up to 2.3 cm long; bracteoles absent. Perianth ovoid, umbilicate at base; segments white with a straw coloured mid-vein, 4–4.5 mm long, ovate, obtuse or subacute, smooth and somewhat shiny, the outer slightly keeled. Stamens with the anthers exserted; filaments ciliate at the base, the outer ones simple, triangular-subulate, the inner ones with the anther-bearing cusp subequal in length to the undivided basal part and very slightly shorter than the lateral cusps; lateral cusps exserted from the perianth; anthers yellow. Style exserted. Capsule ovoid, about 4 mm long; seeds about 3–3.5 mm long, black.

FLOWERING TIME. April–May.

ECOLOGY. Rocky (basaltic) hillsides, 870 m.

DISTRIBUTION. Syria, known only from two collections in the Damascus region.

NOTES. Further collections, including living material, are required in order to ascertain the relationships of this little-known species, in particular with regard to *A.sphaerocephalon*.

62. Allium sphaerocephalon *L.*, Sp. Pl. 1:297(1753); Sims in Curtis's Bot. Mag. 42:t.1764(1815); Sowerby & Smith, Engl. Bot. Suppl. 3:t.2813(1843); Reichenb., Ic. Fl. Germ. 10:24,t.492(1848); Syme, Engl. Bot. 9:t.1533(1869); Regel, All. Monogr. 47(1875)[excl. var. *sardoum*]; Boiss., Fl. Or. 5:236(1882); Fiori & Paol., Ic. Fl. Ital. 80(1898); Halàcsy, Consp. Fl. Graecae 3:246(1904); Coste, Fl. France 3:334(1906); Hegi, Ill. Fl. Mittel-Eur. 2:218(1909); Hayek, Prodr. Fl. Penins. Balcan. 3:42(1932); Vved. in Fl. URSS 4:241(1935), Engl. ed. 4:186(1968); Ross-Craig, Draw. Brit. Pl. 29:t.18(1972); Wilde-Duyfjes in Taxon 22:84(1973); Stearn & Özhatay in Ann. Mus. Goulandris 3:46(1977); Wilde-Duyfjes, Revis. Allium in Africa 44(1977); Stearn in Ann. Mus. Goulandris 4:180(1978); Stearn in Fl. Europaea 5:66(1980); Pastor & Valdes, Revis. Gen. Allium Penins. Iber. & Isl. Bal. 54(1983); Kollmann in Davis, Fl. of Turkey 8:177(1984); Meikle, Fl. of Cyprus 2:1619(1985).

Bulb ovoid to subglobose, 1–2.5 cm diam.; outer tunics pale or yellowish-brown, membranous or coriaceous, sometimes splitting longitudinally into parallel fibres; bulblets present, whitish or yellowish, up to 2 cm long, ovoid, flattened or concave on one side, acuminate, sessile or borne on long stipes well above the bulb beneath the leaf sheaths. Stem (15–)20–90 cm, smooth. Leaves 2–6, shorter than the inflorescence, subterete, caniculate, fistulose, 1–4 mm wide, glabrous, or slightly scabrid-papillose on the margins, sheathing the lower quarter to half of the stem; sheaths smooth. Spathe split into 2(–4) valves, the valves 1–2 cm long, ovate at the base, shortly beaked, shorter than the umbel, persistent and reflexed. Umbel spherical, or broadly ovoid, (1.5–)2–3.5(–6) cm diam., usually very dense, normally wholly floriferous but sometimes also with bulbils present. Pedicels smooth or papillose, unequal, 0.5–2 (–3) cm long; bracteoles few, laciniate, silvery-white. Perianth ovoid-cylindrical to ovoid; segments deep reddish-purple, pink, or white with a green or yellowish keel, 3.5–6 mm long, papillose on the outside, especially on the keel, or smooth, narrowly ovate or lanceolate, obtuse or subacute, often mucronate. Stamens with conspicuously exserted anthers; filaments minutely papillose-dentate at the base, the outer ones triangular-subulate, the inner ones with the anther-bearing cusp equalling to about half as long as the expanded undivided basal part and slightly longer than to slightly shorter than the lateral cusps; lateral cusps exserted; anthers purple or reddish before dehiscence. Style exserted. Capsule broadly ovoid or subglobose, 3.5–4 mm long; seeds 2.5–3 mm long, angular, black. 2n=16,24.

NOTES. *A.sphaerocephalon* is a widespread species, capable of becoming somewhat weedy on account of the freely-produced bulblets. It is fairly readily recognised in its most widespread variant, subsp. *sphaerocephalon*, by the tight umbels of deep purple flowers subtended by persistent reflexed spathe valves. The specific epithet is often spelled *sphaerocephalum* but, as pointed out by Stearn in Ann. Mus. Goulandris 4:181(1978), Linnaeus originally used the ending -*on* which is an equally correct form of spelling.

A.sphaerocephalon exhibits a considerable amount of variation, some of which can be correlated with geographical distribution. Stearn, in Fl. Europaea 5:66(1980), recognised three subspecies, *sphaerocephalon*, *arvense* and *trachypus*, based mainly on flower colour and absence or presence of papillae on the perianth and pedicels. Wilde-Duyfjes (1976) also recognised three subspecies for North Africa, *sphaerocephalon*, *durandoi* and *curtum*. I have seen very little material of subsp. *durandoi* and have accepted this as defined by Wilde-Duyfjes but it is clear that further field work is required to clarify the status of this taxon. However, I regard subsp. *curtum* as representing a distinct species with a more easterly distribution than that of *A.sphaerocephalon*.

A.sphaerocephalon normally has no bulbils present in the umbel, but bulbilliferous variants are known. If it is necessary to refer to these, the name var. *bulbilliferum* Loret & Barrandon is available [see in synonymy below, under subsp. *sphaerocephalon*].

Key to the four subspecies of *A.sphaerocephalon*:

1. Bulblets absent or only 1–2, sessile; Algeria, Tunisia ....subsp. **durandoi**
   Bulblets usually several, stipitate, produced on the neck of the bulb, and sessile at its base; Widespread 2
2. Perianth segments deep purple or pinkish ....subsp. **sphaerocephalon**
   Perianth segments white with a green or yellowish keel 3
3. Perianth and pedicels smooth ....subsp. **arvense**
   Perianth and pedicels papillose ....subsp. **trachypus**

62a subsp. **sphaerocephalon**. Type: Illustration in Micheli, Nova Genera Plantarum t.25,f.2(1729), also represented by a *Micheli* specimen in FI, see Wilde-Duyfjes in Taxon 22:88(1973)).

A.descendens L., Sp. Pl. 1:298(1753) nom. ambig.

Porrum descendens (L.)Reichenb., Fl. Germ. Excurs. 114(1830).

A.sphaerocephalum var. *typicum* Regel, All. Monogr. 46(1875).

A.sphaerocephalum var. *descendens* (L.)Regel, All. Monogr. 47(1875).

A.sphaerocephalum var. *bulbilliferum* Loret & Barrandon, Fl. Montpell. 2:630(1876). Type: France, 'Lamalou; Avene-les-Bains; Pegayrolles-de-l'Escalette et Saint Maurice', *Loret* (?MPU).

A.purpureum Loscos, Trat. Pl. Arag. 1:7(1876). Type: Portugal, near Penarroya, 1865, *Loscos* (isotype COI).

A.sphaerocephaloides Foucaud ex Savatier in Acad. Rochelle, Sci. Nat., Ann. 14(1877):76(1878). Type: not traced.

A.sphaerocephalum var. *bulbiferum* Willk., Ill. Fl. Hisp. 1:157(1885). Type: as for *A.purpureum* Loscos.

A.loscosii Richter, Pl. Eur. 1:199(1890). Type: as for *A.purpureum* Loscos.

?*A.aegaeum* Heldr. & Halàcsy ex Halàcsy in Verh. Zool. Bot. Ges. Wien 49:195(1899). Type: Greece, Naxos Is., *Leonis* 275 (holotype G).

A.sphaerocephalum subsp. *eusphaerocephalum* Briq., Prodr. Fl. Corse 1:288(1910).

A.sphaerocephalon var. *genuinum* Coutinho, Fl. Port. 128(1913).

A.regnieri Maire in Bull. Soc. Hist. Nat. Afr. Nord 22:317(1931). Type: Morocco, Middle Atlas, Kheneg Merzoul, 1 July 1923, *Maire* (holotype MPU, isotype P).

A.margaritaceum var. *robustum* Maire in Bull. Soc. Hist. Nat. Afr. Nord 29:453(1938). Type: Morocco, Great Atlas, Plateau des Lacs, *Maire* (holotype MPU).

ILLUSTRATIONS. Sowerby & Smith, Engl. Bot. Suppl. 3:t.2813(1843); Reichenb., Ic. Fl. Germ. 10:24,t.492(1848); Ross-Craig, Draw. Brit. Pl. 29:t.18(1972); Wilde-Duyfjes, Revis. Allium in Africa 48,f.6(1976); Pastor & Valdes, Revis. Gen. Allium Penins. Iber. & Isl. Bal. 55,f.9(1983); Fl. of Turkey 8:165,f.6,no.17(1984). Plates 10C–D.

Bulblets usually several, stipitate and carried beneath the leaf sheaths above the bulb, sometimes also sessile bulblets present.Perianth segments pink or reddish-purple. 2n=16.

FLOWERING TIME. June–July.

ECOLOGY. Dry stony or rocky slopes, waste places, cultivated fields, roadsides, beaches, scrubland, sea level–1550 m.

DISTRIBUTION. Widespread in Europe, except Scandinavia; Canary Is.; Morocco; Aegean Is.; Cyprus; NW, W & S Turkey; Russia, European part.

62b subsp. **trachypus** *(Boiss. & Spruner)K.Richter*, Pl. Eur. 2:200(1890); Stearn in Ann. Mus. Goulandris 4:181(1978); Stearn in Fl. Europaea 5:66(1980); Kollmann in Davis, Fl. of Turkey 8:178(1984). Type: Greece, 'prope Naupliam Argolidis', *Spruner* (holotype G, photo K!).

A.trachypus Boiss. & Spruner in Boiss., Diagn. Pl. Or. Nov. ser.1,7:114(1846).
A.sphaerocephalum var. *trachypus* (Boiss. & Spruner)Boiss., Fl. Or. 5:236(1882).
A.arvense var. *trachypus* (Boiss. & Spruner)Halacsy, Consp. Fl. Graecae 3:246(1904).

Bulblets usually several, stipitate and carried beneath the leaf sheaths above the bulb, sometimes also sessile bulblets present. Perianth segments white with a green median stripe, scabrid-papillose on the outside; pedicels scabrid-papillose.

FLOWERING TIME. July.
ECOLOGY. Rocky places, limestone cliffs, up to about 1500 m.
DISTRIBUTION. Greece; E Aegean Is., SW Turkey.

62c subsp. **arvense** *(Guss.)Arcangeli*, Comp. Fl. Ital. 702(1882); Stearn in Ann. Mus. Goulandris 4:181(1978); Stearn in Fl. Europaea 5:66(1980); Kollmann in Davis, Fl. of Turkey 8:178(1984). Type: Sicily, Palermo, Val di Mazzara, *Gussone* (holotype NAP).

A.sphaerocephalum var. *viridi-album* Tineo, Cat. Pl. Horti Panorm. 1827:275(1827); Regel, All. Monogr. 47(1875); Boiss., Fl. Or. 5:236(1882). Type: as for *A.arvense* Guss.
A.arvense Guss., Fl. Sic. Prodr. 1:403(1827).
A.sphaerocephalum var. *arvense* (Guss.) Gren. & Godron, Fl. France 3:200(1855); Fiori, Nuova Fl. Anal. Italia 1:208(1923).
A.sphaerocephalon subsp. *viridi-album* (Tineo)K.Richter, Pl. Eur. 1:200(1890); Arcangeli, Comp. Fl. Ital. ed.2:137(1894).

Bulblets usually several, stipitate and carried beneath the leaf sheaths above the bulb, sometimes also sessile bulblets present. Perianth segments white with a green or yellowish median stripe, smooth; pedicels smooth.

FLOWERING TIME. May–July.
ECOLOGY. Cultivated and abandoned fields, scrubland, up to about 1500 m.
DISTRIBUTION. Sicily; Malta; Albania; S Greece, Peloponnese & Cyclades; E Aegean Is., S Turkey. Also recorded for Syria and Lebanon by Mouterde, Nouv. Fl Lib. et Syr. 1:268(1966).

62d subsp. **durandoi** *(Batt. & Trabut) Duyfjes* in Wilde-Duyfjes, Revis. Allium in Africa 50(1976). Type: Algeria, Teniet-el-Haad, *Durando* (holotype G).

A.sphaerocephalum var. *durandoi* Batt. & Trabut, Fl. Alger. Monocot. 155(1884).

ILLUSTRATIONS. Wilde-Duyfjes, Revis. Allium in Africa 51,f.7,1–2(1976).

Bulblets absent or 1–2, sessile. Perianth segments pinkish-purple or white, smooth.

FLOWERING TIME. June–September.
ECOLOGY. Rocky places, hillsides, up to about 1500 m.
DISTRIBUTION. Algeria, Tunisia (after Wilde-Duyfjes, 1983).

63. Allium integerrimum *Zahar.* in Ann. Mus. Goulandris 3:90(1977); Stearn in Ann. Mus. Goulandris 4:187(1978); Fl. Europaea 5:68(1980). Type: Greece, Thessaly, Pieria Distr., Kastroplatanon, foot of Mt. Olympus, 50–100 m, 7 July 1971, *Greuter* 9192 (holotype G, isotype ATH!).

ILLUSTRATION. Ann. Mus. Goulandris 3:90,f.24(1977).

Bulb ovoid, 0.8–1 cm diam.; outer tunics coriaceous; bulblets few, acuminate, or absent. Stem 30–40 cm. Leaves 3–4, much shorter than the inflorescence, linear, fistulose, semiterete, canaliculate in the lower part, 1–1.5 mm wide, sheathing the lower quarter to third of the stem, margins papillose-ciliate. Spathe 2-valved, the valves ovate, shortly acuminate, 5–10 mm long, membranous, persistent. Umbel globose, dense, 1.5–2.5 cm diam. Pedicels unequal, up to 1.2 cm long, smooth; bracteoles absent. Perianth campanulate; segments pinkish-purple, subequal in length, about 4 mm long, the outer ones narrowly elliptic-lanceolate, subacute, keeled, slightly scabrid on the keel, the inner ones ovate, obtuse; Stamens with anthers slightly exserted; outer filaments simple, triangular-subulate, the inner ones ciliolate at the base, with the anther-bearing cusp about a third to half as long as the undivided basal part and subequal to or slightly shorter than the lateral cusps; lateral cusps exserted from the perianth; anthers purplish before dehiscence. Style exserted. Capsule subglobose, about 3 mm long; seeds about 2 mm long.

FLOWERING TIME. June–July.
ECOLOGY. In dry grassy places and in *Quercus coccifera* scrub, 50–600 m.
DISTRIBUTION. NE Greece.
NOTES. *A.integerrimum* is very similar to *A.sphaerocephalon* and may in fact be inseparable from it. Zahariadi, when describing the taxon, provided a comparative table showing the differences between *A.integerrimum* and several other species, one of which was *A.sphaerocephalon*. The critical features were said to be the leaf section, canaliculate in *A.integerrimum* and rounded, non-canaliculate in *A.sphaerocephalon*, the flower colour, pink in the former and reddish-purple in the latter, and the degree of exsertion of the inner filaments. The first two characters are unreliable; *A.sphaerocephalon* does have leaves which are canaliculate at least in the lower part, and the flower colour is variable. Concerning the inner filaments, it was stated that the median cusp is included or slightly exserted in *A.integerrimum* and exserted in *A.sphaerocephalon*; in the former, even when the tip of the median cusp is included, the anthers themselves are exserted, albeit only slightly, so in fact in both species the anthers are exserted, but by differing degrees. From the very limited material of *A.integerrimum* which is available it does seem that the anthers are rather less strongly exserted than in *A.sphaerocephalon*, but one has to bear in mind that this widespread species is very variable and to include such subtleties as this would not extend the degree of variability unduly. However, there is one other character which should be taken into account, and this also concerns the inner filaments. In *A.integerrimum* the median cusp is about one third to half as long as the basal lamina of the filament, whereas in *A.sphaerocephalon* it varies from half as long to equal in length to the basal lamina. It is clear that these features can only be checked satisfactorily in the field so for the present it has been decided to recognise *A.integerrimum*.

64. Allium proponticum *Stearn & N.Özhatay* in Ann. Mus. Goulandris 3:48(1977); Stearn in Ann. Mus. Goulandris 4: 182(1978); Kollmann in Davis, Fl. of Turkey 8:178(1984).

Bulb broadly ovoid, 1.5–2.5 cm diam.; outer tunics greyish, membranous; bulblets few, rather large, about 1.5 cm long, narrowly ovoid, acuminate, pale brown. Stem 40–120(–140) cm, smooth. Leaves about 3, much shorter than the inflorescence, semi-terete, fistulose, canaliculate on the upper (adaxial) side, c. 3(–7) mm wide, scabrid on the margins, sheathing the lower quarter to third of the stem; sheaths smooth. Spathe 2-valved, the valves 0.7–1.3 cm long, ovate at the base, narrowing gradually to an acuminate apex, persistent (but see note below under var. *parviflorum*). Umbel spherical or hemispherical, (2.5–)3–6.5 cm diam., many-flowered. Pedicels slender, papillose towards the apex, subequal, up to 3.5 cm long; bracteoles present, silvery-white, membranous. Perianth narrowly ovoid; segments pinkish-purple or purple, 2.5–4 mm long, scabrid-papillose on the outside, the outer three broadly oblong, keeled, subacute, the inner three ovate, subacute. Stamens with exserted anthers; filaments ciliate at the base, the outer ones triangular-subulate, the inner ones with the anther-bearing cusp slightly shorter than or subequal to the expanded undivided basal part and about two thirds as long as the lateral cusps; lateral cusps well-exserted; anthers pinkish-purple before dehiscence. Style exserted. Capsule broadly ovoid, about 4 mm long.

NOTES. *A.proponticum* is clearly rather similar to *A.sphaerocephalon* but is here maintained as a separate species on the basis of the differences presented by Stearn & Özhatay in Ann. Mus. Goulandris 3:45(1977). It has generally larger, more lax umbels up to 6.5 cm diam. (up to 4 cm in the latter) which in fact look superficially more like those of *A.ampeloprasum* than *A.sphaerocephalon*, although the former belongs to the 'flat-leaved' group of species and is unrelated; additionally, the pedicels in *A.proponticum* are papillose towards the apex (smooth in *A.sphaerocephalon*) and there is a karyotype difference between the two, as explained by N.Özhatay, l.c:47(1977).

Albino variants have been collected in Aydin province in otherwise purple-flowered populations.

Two varieties are recognised:
1. Perianth segments 3.5–4 mm long ....var. **proponticum**
    Perianth segments 2.5–3 mm long ....var. **parviflorum**

64a var. **proponticum**. Type: Turkey, Tekirdağ, near Kumbağ, 16 July 1974, *N.& E.Özhatay* ISTE 30435 (holotype BM, isotype ISTE!).

ILLUSTRATIONS. Ann. Mus. Goulandris 3:49,f.2(1977); Fl. of Turkey 8:165,f.6,no.18(1984).

Perianth segments 3.5–4 mm long. 2n=16.

FLOWERING TIME. (May–)June–July.

ECOLOGY. Rocky slopes, maquis, vineyards, coastal cliffs, beaches, walls, ruins, sea level–600 m.

DISTRIBUTION. W.Turkey (recorded in provinces of Aydin, Balikesir, Izmir, Manisa, Tekirdağ).

64b var. **parviflorum** *Kollmann* in Notes R.B.G.Edinb. 41:264(1983); Kollmann in Davis, Fl. of Turkey 8:179(1984). Type: Turkey, Antalya, Termessos, 20 July 1971, *Shmida & Lev-ari* (holotype HUJ!).

ILLUSTRATIONS. Notes R.B.G. Edinb. 41:264,f.8(1983).

Perianth segments 2.5–3 mm long.

FLOWERING TIME. July.

ECOLOGY. *Quercus*, *Ostrya*, and *Juniperus* forest, hard limestone and marble, 800–900 m.

DISTRIBUTION. S.Turkey (Antalya province).

NOTES. The status of this requires investigation in the field. Although it closely resembles *A.proponticum*, the type specimen of var. *parviflorum*, which is in good condition and collected at anthesis, does not have persistent spathe valves, a notable feature of *A.proponticum*.

65. Allium ebusitanum *Font Quer*, Butll. Inst. Catalana Hist. Nat. 24:145(1924); Stearn in Fl. Europaea 5:67(1980); Pastor & Valdes, Revis. Gen. Allium Penins. Iber. & Isl. Bal. 60(1983). Type: Spain, Balearic Is., Ibiza, Cala de les Torretes, pr. Sta. Agnés, 13 June 1908, *Gross* (lectotype [Herb. Font Quer] BC 62229, isolectotype SEV 44924).

ILLUSTRATIONS. Pastor & Valdes, Revis. Gen. Allium Penins. Iber. & Isl. Bal. 61,f.10(1983).

Bulb ovoid, 1–2 cm diam.; outer tunics membranous to coriaceous, dark brown; bulblets absent or few, ellipsoid, acuminate, dark brown or yellowish, carried on stipes up to 1 cm long. Stem 40–60 cm. Leaves 3–4, much shorter than the inflorescence, linear, fistulose, subterete, 1–2 mm wide, sheathing the lower third of the stem, smooth. Spathe 2-valved, the valves ovate, shortly acuminate, 7–10 mm long, persistent. Umbel hemispherical, lax, 2.5–3 cm diam. Pedicels unequal, 0.5–1.5 cm long, smooth; bracteoles absent. Perianth campanulate; segments pale pink with a darker reddish median vein, 4–4.5 mm long, smooth, the outer ones slightly shorter than the inner, ovate, acute or obtuse and mucronate, the inner ones elliptic, obtuse or truncate. Stamens with anthers exserted; filaments ciliate-denticulate at the base, the outer ones simple, triangular-subulate, the inner ones with the anther-bearing cusp about half as long as the undivided basal part and subequal to or slightly shorter than the lateral cusps; lateral cusps exserted from the perianth; anthers purplish before dehiscence. Style exserted. Capsule subglobose, 3–3.5 mm long; seeds 2 mm long.

FLOWERING TIME. June.

ECOLOGY. Not known.

DISTRIBUTION. Balearic Is., known only from Ibiza.

NOTES. *A.ebusitanum* is very closely allied to *A.sphaerocephalon*. It is said to be distinguishable from the latter by its lax, fewer-flowered umbels and fairly short campanulate, rather than ovoid, perianth.

66. Allium fuscoviolaceum *Fomin* in Monit. Jard. Bot. Tiflis 14:50(1909); Grossh., Fl. Kavk. ed.1,1:202(1928), ed.2,2:122(1940); Vved. in Fl. URSS 4:240(1935), Engl. ed. 4:186(1968); Kollmann in Davis, Fl. of Turkey 8:181(1984). Type: Turkey, Kars prov., 'in monte Askjar-dagh prope Sarykamisch', *S.Michailowsky*(holotype TBI).

ILLUSTRATIONS. Grossh., Fl. Kavk. ed.2,2:t.13,f.6(1940); Fl. of Turkey 8:183,f.7,no.3(1984). Plate 11B.

Bulb ovoid, 0.8–1.5 cm diam.; outer tunics greyish, membranous; bulblets, if present, almost as large as the parent bulb, yellowish. Stem (30–)40–70 cm, smooth. Leaves 3–4, shorter than the inflorescence, linear, subterete, fistulose, canaliculate, 2–3 mm wide, scabrid-papillose on the margin, sometimes only sparsely so towards the tip, sheathing the lower quarter to third of the stem; sheaths smooth. Spathe 2–valved, the valves 5–10 mm long, ovate, shortly beaked, persistent. Umbel spherical or hemispherical, 2–3 cm diam., very dense. Pedicels 5–12 mm long, minutely scabrid-papillose, usually only at the swollen apex; bracteoles present, numerous, white, membranous. Flowers smelling of dung\rotting meat; perianth ellipsoid or campanulate; segments dull purple with a green suffusion, particularly on the midvein, or green with a purplish suffusion, 3–4 mm long, oblong, obtuse or subacute, smooth, the outer prominently keeled. Stamens well-exserted; filaments ciliate in at least the lower half, the outer ones simple, narrowly triangular, the inner ones with the anther-bearing cusp about a third to half as long as the expanded undivided basal part and about equal in length to or longer than the more slender lateral cusps; lateral cusps well-exserted from the perianth; anthers purple before dehiscence. Style exserted by 1.5–2 mm. Capsule globose, 3.5 mm long; seeds 2.5 mm long. 2n=16.

FLOWERING TIME. June–August.

ECOLOGY. Dryish stony hillsides, scree, 830–2510 m.

DISTRIBUTION. E.Turkey (recorded in Ağri, Bitlis, Erzurum, ?Hakkâri, Kars, and Van provinces); Transcaucasia; ?N.Iran.

NOTES. This is closely allied to *A.sphaerocephalon* but differs in having numerous bracteoles and very ciliate filaments. It is also very close to *A.artvinense* but the flower colour is different; these two require further investigation in the field. Interestingly, the flowers have a powerful smell, somewhat dung-like, and very reminiscent of the smell emitted by species of *Biarum* (Araceae) which occur in the same region.

A.fuscoviolaceum var. *compactum* Miscz. ex Grossh., Fl. Kavk. ed.1,1:202(1928) was described from Azerbaijan (Ghanzhinskaya ghuberniya) as having pedicels equal in length to or shorter than the flowers, resulting in a very compact umbel, presumably only about 1 cm in diameter.

67. Allium curtum *Boiss. & Gaill.*, Diagn. Pl. Or. Nov. 2,4:116(1859); Boiss., Fl. Or. 5:245(1882); Post, Fl. of Syria, Palest. & Sinai ed.2,2:638(1933); Feinbr. in Palestine Journ. Bot., Jer. Ser. 3:14(1943); Täckholm & Drar, Fl. of Egypt 3:68(1954); Mouterde, Nouv. Fl. Lib. et Syr. 1:269(1966); Kollmann in Davis, Fl. of Turkey 8:179(1984); Meikle, Fl. Cyprus 2:1621(1985); Kollmann in Rotem 15:69(1985); Kollmann in Feinbr., Fl. Palaestina 4:92(1986). Types: Lebanon, 'in arenosis ultra Abarouh in vicinis Sidonis Syriae', *Gaillardot*(holotype G, isotype P!).

A.sphaerocephalum subsp. *curtum* (Boiss. & Gaill.)Wilde-Duyfjes, Revis. Allium in Africa 52(1976).

ILLUSTRATIONS. Rotem 15:70(1985); Fl. Palaestina 4(plates):126(1986). Plates 11C–D.

Bulb more or less globose, 1.5–2.5 cm diam.; outer tunics pale greyish-brown, membranous; bulblets ovoid, 3–4 mm long, straw coloured. Stem (5–)15–30(–60) cm, smooth, rather stout, wiry. Leaves 2–4, shorter than the inflorescence, subterete, canaliculate, fistulose, 1–2 mm wide, smooth or slightly scabrid, sheathing the lower third of the stem; sheaths smooth. Spathe much shorter than the umbel, split into 2–4 ovate membranous whitish lobes, persistent. Umbel ovoid-conical or ovoid-fastigiate, 1.5–3 cm diam., very dense. Pedicels 2–10 mm long, usually elongating considerably in the fruiting stage and becoming more fastigiate, smooth; bracteoles absent. Perianth ovoid-campanulate or ellipsoid-campanulate; segments with a purple or green median band and pale to white margins, 3–5 mm long, ovate or oblong, obtuse, cucullate, smooth. Stamens with exserted anthers; filaments ciliate at base, the outer ones simple, triangular, the inner ones with the anther-bearing cusp about half as long as the widely expanded undivided basal part and slightly shorter than, equalling, or slightly exceeding, the lateral cusps; lateral cusps exserted from the perianth; anthers purple or yellowish. Style well-exserted, purple or green. Capsule broadly ovoid-conical or subglobose, about 3 mm long; seeds 2–2.5 mm long. 2n=16.

FLOWERING TIME. February–May.

ECOLOGY. Generally in sandy areas, sometimes rocky hillsides, at low altitudes, usually below 100 m.

DISTRIBUTION. S.Turkey, Syria, Lebanon, Israel, ? Jordan, N.Egypt, Cyprus.

NOTES. This species has distinctive ovoid-conical umbels, especially in the bud stage, which become more elongate-fastigiate with age since the central flowers open first and their pedicels rapidly extend during anthesis.

There is variation in flower colour and size and some infraspecific taxa have been described, based on these features, combined with the distribution and ecology. The classification appears to be satisfactory in certain regions but is not so convincing when applied to the whole area of distribution. Clearly further field work is required, particularly in Syria, Lebanon and Cyprus, before this question can be properly resolved. The following is a summary of the infraspecific taxa which have been described, and the characters used to define them.

67a. subsp. **curtum**

Perianth ovoid, 3–3.5 mm long; segments purple with pale or white margins; anthers purple; inner filaments exserted, with the middle cusp longer than the lateral cusps.

67b. subsp. **palaestinum** *Feinbr.* in Pal. Journ. Bot., Jer. Ser. 3:14(1943). Type: Palestine, Jordan Valley, between Wadi Far'a and Beit-Shean, 70 m, 1934, *Eig, Feinbrun & Zohary* s.n.(holotype HUJ).

A.sphaerocephalon var. *viridi-album* auct., non Tineo.

67bb. var. **palaestinum**

Perianth ovoid-oblong, 4–5 mm long; segments green with white margins; anthers yellow; inner filaments exserted, with the middle cusp shorter than the lateral cusps.

67bbb. var. **negevense** *Kollmann* in Feinbr., Fl. Palaestina 4:397(1986). Type: Palestine, Central Negev, Mitzpeh Ramon, opposite radar station, 19 April 1967, *F.Kollmann* 383(holotype HUJ).

Perianth ovoid-oblong, 4–5 mm long; segments pale green with white margins and a purple, greenish-purple or green median vein; anthers purple; inner filaments exserted, with the middle cusp about as long as the lateral cusps.

67c. subsp. **aegyptiacum** *Täckholm & Drar*, Fl. Egypt 3:69(1954). Type: Egypt, Um Zegheiw near El Dikheila, *Drar* s.n.(holotype CAIM).

Perianth ovoid, 3–3.5 mm long; segments purple with paler or white margins; anthers purple; inner filaments included, with the middle cusp longer than the lateral cusps. This is also said to differ from subsp. *curtum* in having larger umbels with the inner erect pedicels 3–4 cm long.

68. Allium regelianum *A.Becker* in Bull. Soc. Nat. Moscow 55,1:146(1880). Type: Ukraine, S. of Volgograd, near Krasnoarmeisk (=Sarepta), *A.Becker*(holotype LE!, isotype K!).

ILLUSTRATIONS. Iljin, in Acta. Hort. Petrop. 40:356,f.178(1929).

Bulb ovoid, about 1 cm diam.; outer tunics grey-brown, membranous; bulblets numerous, brown, ovoid and sharply beaked. Stem 30–60 cm. Leaves 3–4, sheathing the lower third to half of the stem, linear, semi-cylindrical and canaliculate, fistulose, 0.5–2 mm wide, the margin minutely toothed. Spathe persistent, splitting irregularly into 2–4 silvery-transparent valves,shorter than the umbel. Umbel hemispherical or ovoid, dense, 2–3 cm diam. Pedicels 0.5–2 cm long, smooth; bracteoles absent. Perianth cylindric-ovoid; segments purple with a darker nerve, 4–4.5 mm long; oblong-lanceolate with a smooth keel, acute or subacute. Stamens with exserted anthers; filaments glabrous or almost so, the outer ones simple, subulate, the inner ones with the anther-bearing cusp a third to half as long as the basal lamina and short-er than the lateral cusps; anthers violet before dehiscence. Style about 2–3 mm long, exserted. Capsule ovoid-subglobose, 2.5–3 mm long.

FLOWERING TIME. May
ECOLOGY. In salt marshes
DISTRIBUTION. S & E Ukraine.
NOTES. Very similar to *A.sphaerocephalon* and possibly falling within the range of variation of that widespread species. It has a silvery-transparent spathe valve which splits into 2–4 irregular lobes whereas that of *A.sphaerocephalon* is less silvery in appearance and tends to split more regularly into 2 acuminate lobes. Also, the peri-anth segments are smooth in *A.regelianum*, but *A.sphaerocephalon* usually has scabrid segments, at least on the keel.

A.scythicum Zoz. in Ucen. Zap. Khar'kovsk. Derz. Univ. 4:65(1936) may be con-specific with *A.regelianum*. It was described from E. of Herson in the Ukraine and was said to have white or pale lilac flowers in which the lateral cusps of the inner sta-mens were 3 times as long as the central cusp. Field studies in this region are required in order to assess the status of this taxon, and the relationships between it, *A.regelianum* and *A.sphaerocephalon*.

69. Allium pruinatum *Link ex Sprengel*, Syst. Veg. 2:35(1825); Coutinho in Bol. Soc. Brot. 13:96(1896); Coutinho, Fl. Portugal 128(1913), ed.2:152(1939); Stearn in Ann. Mus. Goulandris 4:182(1978); Stearn in Fl. Europaea 5:66(1980); Pastor & Valdes, Revis. Gen. Allium Penins. Iber. & Isl. Bal. 64(1983). Type: Portugal, 'Lusitania', *Link* (holotype B).

A.welwitschii Regel, All. Monogr. 66(1875). Type: Portugal, near Azambuja, 1848-1850, *Welwitsch* 998 (holotype LISU).
A.monspessulanum subsp. *pruinatum* (Link)Richter, Pl. Eur. 202(1890).

ILLUSTRATIONS. Pastor & Valdes, Revis. Gen. Allium Penins. Iber. & Isl. Bal. 65,f.12(1983).

Bulb ovoid, 1–1.5 cm diam.; outer tunics greyish-brown, membranous, splitting lengthways; bulblets subspherical, about 3–5 mm diam., greyish. Stem 20–50 cm, smooth. Leaves 2–3, shorter than the inflorescence, filiform, fistulose, 3–5 mm wide, smooth, sheathing the stem for approximately the lower third; sheaths smooth. Spathe 2-valved, the valves shortly beaked, 0.5–1 cm long, persistent. Umbel hemispherical or ovoid, 1–2 cm diam., many-flowered and dense or with many bulbils and few-flowered, or wholly bulbilliferous. Pedicels unequal, 0.2–1.5 cm long; bracteoles present. Perianth cylindrical; segments pink or purple, the inner often paler than the outer, 4.5–5.5 mm long, narrowly ovate or elliptic-oblong, acute, the outer papillose on the keel and margins. Stamens with the anthers included; filaments ciliate-denticulate, the outer ones simple, triangular-subulate, the inner ones with the anther-bearing cusp about half as long as the expanded undivided basal part and a third to half as long as the lateral cusps; lateral cusps included; anthers yellow. Style included. Capsule subglobose, about 2.5 mm long; seeds about 2 mm long, black. 2n=16.

FLOWERING TIME. May–July.
ECOLOGY. Dry scrub, often in sandy soils, sea level to 700 m.
DISTRIBUTION. C.& S.Portugal.
NOTES. Thought to be endemic to Portugal, but *A.rubrovittatum* var. *occidentale* Rouy ex Willkomm, Suppl. Prodr. Fl. Hisp. 51(1893), based on a specimen collected at Algeciras in S.Spain, is regarded by Stearn(1978) and by Pastor & Valdes(1983) as a synonym of *A.pruinatum*. I have not traced the type, which was collected by Reverchon in 1887.

A bulbilliferous variant of *A.pruinatum* has been described as var. *bulbiferum* Coutinho in Bol. Soc. Brot. 13:96(1896) [Type: Portugal, Cascais, Caparide, *Coutinho* LISU 8303 (lectotype LISU: see Pastor & Valdes 1983:64)]; this name may be used for all variants containing bulbils in the inflorescence.

70. Allium melananthum *Coincy* in Journ. Bot. (Paris) 9:336(1895); Coincy, Eclog. Pl. Hispan. 3:t.11(1897); Stearn in Ann. Mus. Goulandris 4:182(1978); Stearn in Fl. Europaea 5:66(1980); Pastor & Valdes, Revis. Gen. Allium Penins. Iber. & Isl. Bal. 62(1983). Type: Spain, Carthagena, 14 June 1895, *Coincy* (lectotype LY: see Pastor & Valdes l.c.).

ILLUSTRATIONS. Coincy, Eclog. Pl. Hispan. 3:t.11(1897); Pastor & Valdes, Revis. Gen. Allium Penins. Iber. & Isl. Bal. 63,f.11(1983).

Bulb ovoid, 0.8–2 cm diam.; outer tunics pale brown, membranous or subcoriaceous; bulblets few, subspherical, about 5–8 mm diam., yellowish, produced on slender stipes up to 5 cm long. Stem 25–80(–100) cm, smooth. Leaves (2–)4–5, shorter than the inflorescence, filiform, fistulose, about 2 mm wide, smooth, prominently ribbed when dry, sheathing the stem for approximately the lower quarter to third; sheaths smooth. Spathe 2-valved, the valves acuminate, 0.5–1.5 cm long, persistent. Umbel spherical, 1–2(–3.5) cm diam., many-flowered and dense. Pedicels (0.2–)0.5–0.7 (–1.5) cm long, smooth; bracteoles present. Perianth broadly ovoid; segments dark purple or blackish-purple, 2.5–3.5 mm long, strongly papillose on the outside, the outer narrowly ovate, acute or subacute, the inner ones elliptic or oblong, obtuse. Stamens with the anthers slightly exserted; filaments glabrous, the outer ones simple, triangular-subulate, the inner ones with the anther-bearing cusp a third to half as long as the expanded undivided basal part and subequal to or very slightly shorter than the lateral cusps; lateral cusps exserted; anthers purple before dehiscence. Style exserted. Capsule ovoid, about 3 mm long; seeds about 2.5–3.5 mm long, black. 2n=16.

FLOWERING TIME. May–June.
ECOLOGY. Rocky and sandy places, sea level to 400 m.
DISTRIBUTION. SE Spain in the provinces of Almeria and Murcia.
NOTES. A very distinctive species with dark purple-black flowers in dense heads.

71. Allium stylosum *O.Schwarz* in Feddes Repert. Sp. Nov. 36:71(1934); Kollmann in Davis, Fl. of Turkey 8:180(1984). Type: Turkey, Izmir prov., 'in fruticetis semper-virentibus, inprimis *Quercus cocciferae*, collium prope Burnova, solo calcareo, 30–100 m', June 1933, *O.Schwarz* 787(holotype B!).

A.reuterianum Boiss. var. *longicaule* O.Schwarz in Fedde, Repert. Sp. Nov. 36:72(1934). Type: Turkey, Izmir, 'Mimnolos mons [Tahtali Dağ] Smyrnae', 1000–1300 m, *O.Schwarz* 343 (holotype B!).

ILLUSTRATIONS. Fl. of Turkey 8:183,f.7,no.1(1984). Plate 12A.

Bulb ovoid, (0.5–)0.8–1 cm diam.; outer tunics greyish, membranous, fibrous at the neck; bulblets present, pale brown, acuminate, stipitate, produced beneath the sheaths on the underground part of the stem, often well above the parent bulb. Stem 35–60(–80) cm, smooth. Leaves usually 3–4, shorter than the inflorescence, linear, sub-terete, fistulose, canaliculate, at least in the lower part, 1–2 mm wide, slightly scabrid-papillose on the margin, sheathing the lower third to half of the stem; sheaths smooth or minutely scabrid. Spathe 2-valved, the valves ovate, shortly mucronate, persistent. Umbel spherical, 2–3 cm diam., very dense. Pedicels 6–10 mm long, smooth or slightly scabrid-papillose, usually only at the extreme apex; bracteoles usually absent. Perianth cylindrical or ovoid; segments deep reddish-purple (rarely white), 4–5(–6) mm long, minutely scabrid outside, obtuse or rounded, sometimes bluntly mucronate, the outer broadly ovate, the inner slightly longer, broadly elliptic. Stamens exserted; filaments minutely ciliate at base, the outer ones simple, triangular, the inner ones with the anther-bearing cusp about a half to two thirds as long as the expanded undivided basal part and between a half and two thirds as long as the lateral cusps; lat-eral cusps well-exserted from the perianth; anthers purple before dehiscence. Style strongly exserted by 2–4 mm. Capsule globose, 5 mm long. 2n=16,32.

FLOWERING TIME. May–June(–July).

ECOLOGY. Dryish habitats including stony hillsides on limestone formations, scrub, sparse coniferous woods, steppe, fields, 100–1800 m.

DISTRIBUTION. W, SW & C.Turkey (recorded in Ankara, Antalya, Çankiri, Denizli, Isparta, Izmir, Kayseri, Konya, Kütahya, Muğla and Üsak provinces).

NOTES. This is closely allied to *A.sphaerocephalon* and differs mainly in the length of the stamens in relation to the perianth segments. In *A.sphaerocephalon* the outer stamens are longer than the perianth segments, and the middle cusp of the inner fila-ments is slightly shorter to slightly longer than the two lateral cusps; in *A.stylosum* the outer stamens are about the same length as the segments and the middle cusp of the inner stamens is only about a half to two thirds as long as the lateral cusps. Additionally, the leaves are normally sparsely scabrid-papillose on the margins in *A.stylosum* and usually glabrous in *A.sphaerocephalon*, although they may occasion-ally be very slightly papillose. *A.reuterianum* var. *longicaule* O.Schwarz was distin-guished from *A.reuterianum* on the basis of its taller stems but it appears to be better placed with *A.stylosum*.

It has been noted (Kollmann, loc. cit.) that some specimens from Isparta and Konya provinces (N. end of Hoyran Gölü, 950 m, *Runem. & Wendelbo*, & Sultan Dağ, Cankarturan, 1400 m, *Sorger* 68–46–31) have white flowers with purple filaments, anthers and style. These populations should be investigated in the field to ascertain the degree of uniformity of the features and whether there is correlation with any other factors.

72. Allium eldivanense *N.Özhatay* in Notes R.B.G. Edinb. 44:147(1986); N.Özhatay in Davis, Fl. of Turkey 10:221(1989). Type: Turkey, Çankiri Prov., Eldivan Dağ, above Bakirli, 18 July 1976, *A. & T.Baytop* ISTE 35244(holotype ISTE!).

ILLUSTRATIONS. Notes R.B.G. Edinb. 44:148,t.1(1986).

Bulb ovoid, 1–1.5cm diam.; outer tunics pale brown, membranous; bulblets not evident. Stem (20–)25–30(–35) cm, smooth, usually flexuose and suffused purple at the base. Leaves 2–3, shorter than the inflorescence, linear, subterete, fistulose, slightly canaliculate, at least in the lower part, 1–2 mm wide, glabrous, sheathing the lower quarter to third of the stem; sheaths smooth, with prominent veins. Spathe 2–3-valved, the valves ovate, acute or very shortly beaked, 5–7 mm long, persistent. Umbel globose, 1.5–2.5 cm diam., dense. Pedicels up to 7 mm long, smooth; bracteoles absent. Perianth ellipsoid-campanulate; segments pinkish-purple with a darker purple or green mid-vein, 4–5 mm long, the outer ones slightly shorter than the inner, ovate, subacute, cymbiform, slightly scabrid on the keel, the inner oblong-lanceolate, truncate or emarginate, smooth. Stamens with long-exserted anthers; filaments minutely ciliate at base, the outer ones simple, triangular-subulate, the inner ones with the anther-bearing cusp about a third as long as the expanded undivided basal part and half as long as the lateral cusps; lateral cusps well-exserted from the perianth; anthers purple before dehiscence. Style strongly exserted, 4–5 mm long. Capsule ovoid, 3–4 mm long; seeds not seen. 2n=16.

FLOWERING TIME. July.
ECOLOGY. Dryish open rocky or stony places, 1550–1850 m.
DISTRIBUTION. N.Turkey (recorded only from the type locality in Çankiri province).
NOTES. This is related to *A.stylosum* O.Schwarz but differs in having pinkish-purple flowers (not deep reddish-purple) which have long-exserted filaments; it also has strong affinities with *A.sphaerocephalon* but has inner filaments with the median cusp much shorter than the two lateral cusps. It is known at present only from the type locality.

73. Allium reuterianum *Boiss.*, Diagn. Pl. Or. Nov. 1,5:60(1844); Boiss., Fl. Or. 5:238(1882); Meikle in Kew Bull. 9:178(1954); Kollmann in Davis, Fl. of Turkey 8:181(1984). Type: Turkey, Manisa prov., 'in ipso cacumine excelsiori orientali montis Sipyli(Manisa Dağ) supra Magnesiam', July 1842, *Boissier*(holotype G).

ILLUSTRATIONS. Kew Bull. 9:179,f.11(1954); Fl. of Turkey 8:183,f.7,no.2(1984).

Bulb ovoid, 1–1.5 cm diam.; outer tunics blackish-brown, membranous; bulblets absent. Stem (5–)10–13(–15) cm, smooth or minutely papillose-scabrid, flexuose. Leaves usually 2, equalling or longer than the inflorescence, narrowly linear, fistulose, canaliculate, 1–2 mm wide, scabrid-papillose on the margin, sheathing the lower half to three quarters of the stem; sheaths smooth or minutely scabrid. Spathe 2-valved, the valves ovate-rounded, shortly mucronate, persistent. Umbel spherical, 1–1.5 cm diam., dense. Pedicels 3–5 mm long, scabrid-papillose, sometimes only at the extreme apex; bracteoles absent, or sometimes a few present adjacent to the outer flowers, white, membranous, usually associated with sterile flowers. Perianth shortly campanulate; segments deep reddish-purple or crimson, 3–3.5 mm long, ovate-oblong, obtuse or subacute, mucronate, the outer ones densely scabrid all over the outer surface and coarsely so on the keel. Stamens exserted; filaments slightly ciliate at base, the outer ones simple, triangular, the inner ones with the anther-bearing cusp about a third as long as the widely expanded undivided basal part and about three quarters as long as the lateral cusps; lateral cusps well-exserted from the perianth; anthers purple before dehiscence. Style strongly exserted by 1–4 mm. Capsule ovoid, 2–2.5 mm long. 2n=16.

FLOWERING TIME. July–August.
ECOLOGY. Stony mountain slopes on limestone formations, 1800–2100 m.
DISTRIBUTION. W & SW Turkey(recorded in Antalya, Denizli, Isparta, Izmir, Manisa and Muğla provinces); E Aegean Is.(Khios, Samos).
NOTES. This is closely allied to *A.stylosum* and could perhaps be regarded as a diminutive high altitude variant of the latter. Further field studies are desirable. *A.reuterianum* var. *longicaule* O.Schwarz, described from Izmir, Tahtali Dağ, was said to differ mainly from *A.reuterianum* in being taller; it is probably best referred to *A.stylosum*.

74. Allium artvinense *Miscz. ex Grossh.*, Fl. Kavk. ed.1,1:203(1928), ed.2,2:124(1940); Kollmann in Davis, Fl. of Turkey 8:185(1984). Type: Turkey, Çoruh (Artvin) prov., between Zetilet and Skalimer on the Artvin to Ardahan road, 26 July 1911, *Vvedensky* 3760 (holotype LE).

ILLUSTRATIONS. Fl. of Turkey 8:183,f.7,no.7(1984). Plate 12B.

Bulb ovoid, 1–2 cm diam.; outer tunics greyish-brown, membranous; bulblets present, elongate, yellowish. Stem 30–120 cm, smooth, slightly ribbed. Leaves 3–5, shorter than the inflorescence, linear, subterete, fistulose, canaliculate, 2–3 mm wide, glabrous or scabrid in the lower part, sheathing the lower half to two thirds of the stem; sheaths smooth or scabrid, conspicuously veined, slightly ribbed. Spathe 2–3-valved, the valves ovate with acute or shortly acuminate tips, 1–1.2 cm long, persistent. Umbel spherical or hemispherical, 2.5–5 cm diam. Pedicels 1.3–2 cm long, minutely scabrid-papillose, usually only at the apex; [the type specimen also has many short-pedicelled sterile flowers, but this is probably not a normal characteristic of the species]; bracteoles present, white, membranous, narrowly linear. Perianth oblong; segments white or green, sometimes faintly flushed pink, with a green median vein on the outside, 3–5 mm long, smooth, oblong or ovate, acute or subacute, prominently keeled. Stamens slightly exserted; filaments ciliate in the lower half, the outer ones simple, narrowly triangular, the inner ones with the anther-bearing cusp about a third to a half as long as the expanded undivided basal part and slightly shorter to slightly longer than the lateral cusps; lateral cusps exserted from the perianth; anthers purple before dehiscence. Style 3–4 mm, exserted, white. Capsule subglobose, 3–5 mm long; seeds black, 3mm long. 2n=16.

FLOWERING TIME. (June–)July–August.
ECOLOGY. Dryish rocky hillsides, screes, steppe, (860–)1650–2410 m.
DISTRIBUTION. E.Turkey (recorded in provinces of Çoruh (Artvin), Erzincan, Erzurum, Giresun, Gümüşane, Sivas and Van); Armenia, in the Sevan area.
NOTES. Further studies are desirable since this has not been adequately assessed in the field. It is apparently closely related to *A.fuscoviolaceum*, from which it differs most obviously in flower colour, and is also allied to *A.nevsehirense* but has, in general, larger, more lax, umbels.

75. Allium makmelianum *Post*, Pl. Post. 3:18(1892); Post, Fl. of Syr., Pal. & Sinai 788(1896); Feinbr. in Pal. Journ. Bot., Jer. Ser. 3:12(1943); ed.2,2:635(1933); Thiébaut, Fl. Lib.-Syr. 3:179(1953); Mouterde, Nouv. Fl. Lib. et Syr. 1:268(1966).

Type: Lebanon, Rijal-el-'Asherah, 22 July 1891, *Post* (holotype BEI)[the Keeper of the Herbarium confirmed that the type is in the Post collection but at the time of writing it was impossible to send this on loan].

ILLUSTRATIONS. Mouterde, Nouv. Fl. Lib. et Syr. 1(Atlas):pl.83(1966).

Bulb ovoid, about 1.5 cm diam.; outer tunics blackish, membranous; bulblets not apparent. Stem 5–15 cm, densely and minutely scabrid-papillose on the prominent veins. Leaves 2–3, green at anthesis, exceeding the infloresence, sheathing to more than half the stem, linear, semiterete, fistulose, 1–2 mm wide, with prominent ribs, at least when dry, densely scabrid-papillose on all the veins; sheaths scabrid-papillose. Spathe 2-valved, the valves about 1 cm long, ovate at the base, shortly beaked, scarious, densely scabrid-papillose, persistent. Umbel globose, 1.5–2.5 cm diam., dense. Pedicels up to 0.8 cm long; bracteoles few, small, scarious. Perianth ovoid-urceolate, umbilicate at base; segments white with a reddish mid-vein, 6–7 mm long, ovate-lanceolate, acuminate or mucronate, minutely papillose on the outside. Stamens with the anthers included; filaments glabrous, the outer ones simple, triangular-subulate, the inner ones with the anther-bearing cusp a quarter to a third as long as the very widely expanded undivided basal part and a sixth to a quarter as long as the lateral cusps; lateral cusps shortly exserted from the perianth; anthers reddish-brown. Style included. Capsule ovoid, much shorter than the perianth (according to Mouterde).

FLOWERING TIME. June–July.
ECOLOGY. Gravelly summits, rocky places, near snow-line.
DISTRIBUTION. C & N Lebanon (Lebanon and Anti-Lebanon Mts); adjacent Syria (Yabroud & Jabal Ma'loula).
NOTES. A very distinct species, unusual in its habitat preference in mountains near the snow-line. The dwarf habit, relatively large flowers and scabrid-papillose stem, leaves, spathe valves and perianth segments, make it instantly recognisable.

76. Allium kotschyi *Boiss.*, Diagn. Pl. Or. Nov. 1,7:117(1846); Boiss., Fl. Or. 5:234(1882); Wendelbo in Rechinger, Fl. Iranica 76:46(1971). Type: Iran, Fars prov., 'in glareosis mobilibus Kuh-e Dinar', *Kotschy* 777 (holotype G, isotypes GB,K!).

ILLUSTRATIONS. Fl. Iranica 76:t.5,f.58(1971); Matine, Contr. Etude Fam. Alliaceae en Iran pl.2,no.1(1976).

Bulb ovoid, 1–1.5 cm diam.; outer tunics greyish-brown, membranous, splitting into parallel fibres, produced at the apex into a neck up to 2 cm long; bulblets not apparent. Stem 5–12 cm, slightly ribbed and tuberculate. Leaf solitary, exceeding the inflorescence, linear, subterete, fistulose, apparently caniculate, ribbed, about 1–2 mm wide, smooth, sheathing the lower half of the stem; sheaths smooth. Spathe with 1(–2 if split) ovate, acuminate valves, up to 5 mm long, persistent. Umbel hemispherical, with up to 20 flowers, about 2–2.5 cm diam. Pedicels 5–9 mm long, smooth, thickened at the apex; bracteoles present, white, membranous. Perianth campanulate; segments white with a dark purplish median vein, 5–6.5 mm long, lanceolate, acute, smooth, the outer slightly longer than the inner, margins minutely erose-dentate. Stamens included; filaments smooth, the outer ones simple, narrowly triangular, the inner ones with the anther-bearing cusp only about one sixth as long as the broad undivided basal part and one third as long as the lateral cusps; lateral cusps included well within the perianth; anthers yellow. Style not exserted. Capsule subglobose, about 3 mm long; seeds not seen.

FLOWERING TIME. July–August

ECOLOGY. Gravelly slopes in mountains.

DISTRIBUTION. Iran, in Fars province.

NOTES. Although, in Flora Iranica, *A.kotschyi* is described as having non-fistulose leaves, it is clear from the isotype specimen at Kew that they are in fact semi-terete and hollow. As far as I can ascertain, the species has not been recollected since the original gathering by Kotschy in 1845.

77. Allium lehmannianum *Merckl.* in Mem. Acad. Petersb. 7:509(1851); Regel, Monogr. All. 73(1875); Regel, Fl. Turkest. 1:41(1876); Boiss., Fl. Or. 5:234(1884)[excl. var. *bungei* Boiss.]; Vved. in Fl. URSS 4:233(1935), Engl. ed. 4:178(1968); Vved. in Fl. Uzbek. 1:450(1941); Pavl. & Poljak. in Fl. Kazakh. 2:183(1958); Vved. in Consp. Fl. Asiae Mediae 2:77(1971). Type: C.Asia, Kyzyl Kum Desert, between rivers Kuwan Darya and Syr Darya, *Lehmann* (holotype LE, isotype K!,P!).

ILLUSTRATIONS. Regel, Fl. Turkest. 1:t.6,f.8,9,10(1876).

Bulb ovoid, 0.75–1.5 cm diam.; outer tunics greyish-brown or reddish-brown, membranous or coriaceous; bulblets absent. Stem 5–8(–15) cm, smooth. Leaves 2–3, longer than the inflorescence, filiform, fistulose, about 1 mm wide, smooth, clustered together and sheathing only about the lower eighth of the stem; sheaths smooth. Spathe 2–3-valved, the valves ovate-rounded, acuminate, not exceeding the umbel, persistent. Umbel spherical or hemispherical, 2–2.5 cm diam., few-flowered, lax. Pedicels 5–12 mm long, smooth; bracteoles present, silvery-white, membranous. Perianth campanulate; segments with white margins and a purple median vein, 6–7 mm long, lanceolate or oblong, acute or subacute, smooth. Stamens included; filaments glabrous, the outer ones simple, lanceolate-triangular, the inner ones with the anther-bearing cusp about a third as long as the expanded undivided basal part and about a half to two thirds as long as the lateral cusps; lateral cusps included; anthers purplish(?) before dehiscence. Style included. Capsule suborbicular, about 4 mm long.

FLOWERING TIME. May–June.

ECOLOGY. Sandy soils, desert.

DISTRIBUTION. C.Asia, the Aral-Caspian and Kyzyl Kum deserts.

NOTES. *A.lehmannianum* is one of several species from Central Asia which have a relatively dwarf stature with leaves overtopping the inflorescences. *A.borszczowii* and *A.brevidens*, although rather similar, differ from it in having reticulated bulb tunics; *A.ferganicum* has membranous-coriaceous tunics like those of *A.lehmannianum* but can be distinguished by the shorter perianth segments and different proportions of the parts of the inner filaments.

78. Allium mareoticum *Bornm.* & *Gauba* in Fedde, Repert. Sp. Nov. 31:396(1933); Täckholm & Drar, Fl. of Egypt 3:70(1954). Type: Egypt, 'in Aegypti inferioris territorio Marmarica ad El Omaied', 23 March 1930 (fl.), 23 July 1930 (fr.), *Gauba* (holotype TEH?).

Bulb oblong-ovoid; outer tunics pale brown, membranous, extended into a short neck up to 3.5 cm long; bulblets few, about 1 cm long, brown, produced on the neck of the bulb beneath the leaf sheaths. Stem 12–25 cm. Leaves 2–4, much shorter than the inflorescence, sheathing the lower third to quarter of the stem, linear-filiform, semiterete, fistulose, canaliculate, 1–2 mm wide, with prominent ribs, at least when dry, glabrous; sheaths glabrous. Spathe caducous, ['apparently 2-valved' according to original descr.]. Umbel globose, (1.5–)2–3 cm long, dense. Pedicels unequal, up to 1.2 cm long; bracteoles present, rather conspicuous, somewhat laciniate, white-scarious. Perianth campanulate, umbilicate at base; segments white with a pinkish or green mid-vein, 3 mm long, oblong, obtuse and minutely apiculate, the outer scabrid on the keel. Stamens with the anthers exserted; filaments glabrous, the outer ones simple, narrowly triangular, the inner ones with the anther-bearing cusp about a quarter as long as the expanded undivided basal part and a third as long as the lateral cusps; lateral cusps well-exserted from the perianth; anthers apparently purplish. Style included or very shortly exserted. Capsule ovoid-globose, about 2.5–3 mm long; seeds black.

FLOWERING TIME. (March–)April–June.
ECOLOGY. Sandy and rocky places, on calcareous hills.
DISTRIBUTION. N.Egypt.
NOTES. This is apparently a very rare plant in Egypt, much of its habitat having disappeared due to urbanisation and agriculture. It is of small stature with very narrow, short leaves and has early-deciduous spathe valves; this last feature distinguishes it from *A.curtum* and *A.sphaerocephalon* and their allies.

79. Allium ferganicum *Vved.* in Not. Syst. Herb. Hort. Bot. Petrop. 5:90(1924); Vved. in Fl. URSS 4:229(1935), Engl. ed. 4:178(1968); Vved. in Fl. Uzbek. 1:450(1941); Kaschtsch. & E.Nikit. in Fl. Kirgiz. 3:87(1951); Vved. in Fl. Tadjik. 2:331(1963); Vved. in Consp. Fl. Asiae Mediae 2:77(1971). Type: C.Asia, Pamir-Alai Mts., S of Kokand, *Werner* (syntype LE); Rushtan to Chongoru, *Tukaeva* 368 (syntype LE); Skobelev, Ak-Tubi, *A.& P.Archangelsky* (syntype LE).

A.lehmannianum var. *kokanicum* Regel in Acta Horti Petrop. 10:304(1887). Type: C.Asia, 'in Kokania prope Osch ad rupem Suleiman', *A.Regel* (holotype LE).

Bulb ovoid, 0.5–1 cm diam.; outer tunics greyish, membranous; bulblets few, rather large, yellowish. Stem 10–20 cm, smooth. Leaves 2–4, longer than the inflorescence, filiform, fistulose, about 1 mm wide, smooth, clustered together near the base of the stem; sheaths smooth. Spathe not exceeding the umbel, shortly beaked, caducous. Umbel spherical or hemispherical, 2–4 cm diam., dense. Pedicels 0.6–1.5 mm long; bracteoles present. Perianth campanulate; segments pink with a purple median vein, 4–6 mm long, oblong-lanceolate, acute, smooth(?). Stamens included; filaments glabrous, the outer ones simple, triangular-subulate, the inner ones with the anther-bearing cusp two thirds as long as to subequal to the expanded undivided basal part and equalling or exceeding the lateral cusps; lateral cusps included; anther colour unknown. Style included. Capsule suborbicular, about 4 mm long.

FLOWERING TIME. May.
ECOLOGY. Desert foothills.
DISTRIBUTION. C.Asia, in the foothills of the Pamir-Alai range.
NOTES. *A.ferganicum* may be likened to *A.lehmannianum* but can be distinguished by having shorter perianth segments and different proportions of the parts of the inner filaments, the median cusp being longer than the lateral cusps (half to two thirds as long as in *A.lehmannianum*) and equal to or two thirds as long as the basal lamina (a third as long as in *A.lehmannianum*). Additionally, *A.ferganicum* is a taller plant with more densely-flowered umbels of smaller flowers.

80. A.guttatum *Steven* in Mém. Soc. Nat. Moscou 2:173(1809); Vved. in Fl. URSS 4:238(1935), Engl. ed. 4:184(1968); Wilde-Duyfjes, Revis. *Allium* in Africa 34(1977); Stearn in Ann. Mus. Goulandris 4:184(1978); Stearn in Fl. Europaea 5:67(1980); Kollmann in Davis, Fl. of Turkey 8:187(1984).

Bulb ovoid, 1–2 cm diam.; outer tunics brown, grey or blackish, membranous or coriaceous, sometimes splitting into parallel fibres; bulblets absent or few, narrowly ovoid, acuminate, yellowish or pale brown, smooth. Stem 10–100 cm, smooth. Leaves 2–5(–6), shorter than, equalling or exceeding the inflorescence, linear-fili-form, fistulose, usually flattened on the upper (adaxial) side, 1–3 mm wide, smooth or scabrid-papillose on the veins, sheathing the lower third to two thirds of the stem; sheaths smooth. Spathe 1-valved, up to 4(–6) cm long, broadly ovate at the base, nar-rowing abruptly to a long beak, caducous. Umbel spherical, ovoid or hemispherical, 1.5–5 cm diam. Pedicels smooth, unequal, up to 3 cm long, often the inner ones quickly becoming erect in fruit while the outer ones are still patent or deflexed, thus giving the impression of a 'two-tiered' umbel; bracteoles present, sometimes united into a membranous silvery-white involucre surrounding the base of the umbel, some-times laciniate into several to many filiform divisions. Perianth cylindrical-campan-ulate to narrowly ellipsoid; segments white, pink or purple with a green, pinkish or purplish median vein or blotch, 2–4(–4.5) mm long, smooth, subequal or the outer slightly longer than the inner, the outer three narrowly oblong, keeled, obtuse or sub-acute, the inner three oblong or narrowly obovate, truncate-rounded or obtuse, some-times mucronate. Stamens with exserted anthers; filaments glabrous or minutely ciliate-dentate, the outer ones triangular-subulate, the inner ones with the anther-bear-ing cusp slightly shorter to slightly longer than the expanded undivided basal part and about one third to half as long as the lateral cusps; lateral cusps exserted; anthers yel-low or reddish-purple before dehiscence. Style exserted. Capsule subglobose, 3–4 mm long; seeds 2–3 mm long, angular-triquetrous, black.

NOTES. *A.guttatum* is related to *A.amethystinum* and the two species share the char-acteristic of often having a 'double-umbel' effect caused by the inner, fertilised, flow-ers rapidly producing capsules and standing up in a cluster on elongated erect pedicels above the rest of the flowers in the umbel; this feature is much more developed in *A.amethystinum* than in *A.guttatum*. In general, the former is a larger more robust plant with larger umbels of purple flowers which usually have the inner segments 3.5–6 mm long (usually 2–4 mm in *A.guttatum*, except in the white-flowered subsp. *dilatatum* from Crete); also the leaves of *A.amethystinum* are up to 8 mm wide and very deeply canaliculate, often with a pale median stripe, whereas in *A.guttatum* they are filiform, at most 3 mm wide and only flattened, or very slightly canaliculate, on the upper surface.

Although this taxon has frequently been referred to as *A.margaritaceum* Smith (1809), the use of this epithet is invalidated by *A.margaritaceum* Moench (1802), as pointed out by Stearn [Ann. Mus. Goulandris 4:185(1978)]; *A.guttatum* Steven (pub-lished later on in 1809) therefore becomes the correct name for this widespread and familiar species.

The morphological variation in *A.guttatum* has a certain amount of geographical significance and it is appropriate to utilise the rank of subspecies to define at least some of the variants. Stearn, in Ann. Mus. Goulandris 4:184(1978), identified three subspecies and a fourth, subsp. *dilatatum*, is added here, based on *A.dilatatum* Zahariadi. The characters used by Zahariadi to define this Cretan endemic are uncon-vincing at specific level and indeed, most of the features fall within the accepted vari-ation of *A.guttatum* subsp. *sardoum*.

Key to the subspecies of *A.guttatum*:

1. Perianth segments pink to purple ....subsp. **dalmaticum**
   Perianth segments white with a green, purplish or brownish median stripe or blotch 2
2. Perianth segments each with a suborbicular or oblong green, purple or brown blotch ....subsp. **guttatum**
   Perianth segments with a green or pinkish-purple median stripe 3
3. Inner perianth segments 5 mm long ....subsp. **dilatatum**
   Inner perianth segments usually 2–4 mm long ....subsp. **sardoum**

80a. subsp. **guttatum**. Type: 'Circa colonias Svevicas districtus Odessani', *Steven* (holotype H, isotypes FI photo!,K!,LE,WAG!).

A.margaritaceum var. *guttatum* (Steven)Gay in Ann. Sci. Nat. 3, Bot. 8:223(1847); Boiss., Fl. Or. 5:240(1882); Halàcsy, Consp. Fl. Graecae 3:245(1904); Meikle, Fl. of Cyprus 2:1625(1985).
? *A.sphaerocephalum* subsp. *rumelicum* Formanek in Verh. Naturf. Ver. Brünn. 36:41(1898). Type: unknown.

ILLUSTRATIONS. Mém. Soc. Nat. Moscou 2:173,t.11,f.1(1809); Fl. of Turkey 8:183,f.7,no.10(1984). Plate 12C.

Perianth segments whitish with a suborbicular or oblong purple, brown or green median blotch. 2n=16.

FLOWERING TIME. July–August.
ECOLOGY. Rocky hillsides, screes, sandy coastal areas, open coniferous woods, scrub, roadsides, sea level–2400 m.
DISTRIBUTION. Bulgaria; Roumania; E.Aegean Is.; NW, C & SE Turkey; SW Russia (European part); Crimea.

80b. subsp. **sardoum** *(Moris)Stearn* in Ann. Mus. Goulandris 4:184(1978); Stearn in Fl. Europaea 5:67(1980); Pastor & Valdes, Revis. Gen. Allium Penins. Iber. & Isl. Bal. 66(1983). Type: Italy, Sardinia, 'in pascuis aridis circa Mandas' (holotype SASSA or TO ?). (See note, p.164).

A.margaritaceum Smith in Sibth. & Smith, Fl. Graecae Prodr. 1:224(1809); Fl. Graeca 4:14(1823); Boiss., Fl. Or. 5:239(1882); Halàcsy, Consp. Fl. Graecae 3:245(1904); Meikle, Fl. of Cyprus 2:1624(1985). Syntypes: Turkey, Greece and Cyprus, 'circa Bursam Bithyniae; etiam in monte Athö, insulisque Naxo, Cypro et Cimalo' *Sibthorp* (OXF)!
A.sardoum Moris, Stirp. Sard. 2:10(1828).
?*A.lineare* Ten., Fl. Nap. 1:164(1831). Type: Italy, Calabria, *Gussone* (holotype ?NAP or FI).
A.densiflorum Hampe in Flora (Regensburg) 25:80(1842). Type: Greece, Attica (not traced).
A.frivaldszkyanum Kunze in Linnaea 16:311(1842). Type: Macedonia, *Frivaldszky* (holotype ?BP).
A.margaritaceum var. *tenorii* Parl., Fl. Ital. 2:569(1852). Type: Italy, Calabria, Sila, *Tenore* (holotype NAP).
A.gaditanum Pérez-Lara in Willk., Ill. Fl. Hisp. 1:81(1882). Type: Spain, 'in dumosis dehesa del Malduerme, urbis Jerez', 26 July 1879, *Pérez-Lara* MAF33700(lectotype MAF).

A.sphaerocephalum subsp. *sardoum* (Moris)K.Richter, Pl. Europ. 1:200(1890).
A.margaritaceum subsp. *tenorii* (Parl.)K.Richter, Pl. Europ. 1:200(1890).
A.involucratum Welw. ex Coutinho in Bol. Soc. Broter. 13:98(1896). Type: Portugal,
    Serra d'Arrabida, *Welwitsch* LISU 65728 (lectotype LISU).
A.margaritaceum var. *typicum* Regel, All. Monogr. 50(1875).
A.sphaerocephalon var. *sardoum* (Moris)Regel, All. Monogr. 47(1875); Fiori in Fiori
    & Paol., Fl. Anal. Ital. 1:196(1896).
A.confusum Halàcsy, Consp. Fl. Graec. 3:244(1904). Syntypes: Greece, Thessaly,
    near Malakasi, Korona Monastery, *Haussknecht*; Karava, *Sintenis* (not traced).
A.margaritaceum var. *faurei* Maire in Bull. Soc. Hist. Nat. Afr. Nord. 28:381(1937).
    Type: Morocco, 'colline de Taghit', 17 May 1933, *Faure* (holotype MPU, isotype
    P).
A.negrianum Maire & Weiller in Bull. Soc. Hist. Nat. Afr. Nord 30:304(1939). Type:
    Libya, Cyrenaica, Ouadi Kouf, *Maire* 1739 (holotype MPU).
A.margaritaceum var. *battandieri* Maire & Weiller in Bull. Soc. Hist. Nat. Afr. Nord.
    34:191(1945). Type: Algeria, between Oussengh and Itema, *Battandier* (holotype
    MPU, isotype P).

ILLUSTRATIONS. Sibth. & Sm., Fl. Graeca 4:t.315(1823); Wilde-Duyfjes, Revis.
Allium in Africa 36,f.4(1976)[as *A.guttatum*]; Fl. of Turkey 8:183,f.7,no.11(1984);
Pastor & Valdes, Revis. Gen. Allium Penins. Iber. & Isl. Bal. 67,f.13(1983). Plate
12D.

Perianth segments white with a green or rarely pinkish median line or stripe, the
inner three usually 2–4 mm long. 2n=16,24,32,40.

FLOWERING TIME. July–August.
ECOLOGY. Rocky places, maquis, open coniferous woods, vineyards, sea
level–2200 m.
DISTRIBUTION. Portugal, east through Mediterranean region to Balkans; W,SW,S.&
SE.Turkey; N.Africa.

80c. subsp. **dalmaticum** *(A.Kern. ex Janchen)Stearn* in Ann. Mus. Goulandris
4:184(1978); Stearn in Fl. Europaea. 5:67(1980); Kollmann in Davis, Fl. of Turkey
8:188(1984). Type: former Yugoslavia, 'Ledenica in der Krivosije', 650 m,
July–August 1915, *Knoll* (lectotype WU!, isolectotype K!: see Stearn in Ann. Mus.
Goulandris 4:185[1978]).

A.dalmaticum A.Kerner ex Janchen in Osterr. Bot. Zeitschr. 69:242(1920); Hayek,
    Prodr. Fl. Penins. Balcan. 3:41(1932).
?*A.margaritaceum* var. *rubellum* Boiss., Fl. Or. 5:240(1882). Type: Bulgaria, 'in mon-
    tanis Hami prope Kalofer Thraciae', *Pichler* (holotype G).
?*A.rilaense* Panov in Comptes Rend. Acad. Bulg. Sci. 26:1227,t.1,f.1(1973). Type:
    Bulgaria, Mt. Rila, 12 July 1964, *P.Panov* SOM-125851 (SO)

ILLUSTRATIONS. Reichenb., Ic. Fl. Germ. 10:t.491(1848)[as *A.margaritaceum*].

Perianth segments pink to purple, not conspicuously blotched or striped. 2n=24,32.

FLOWERING TIME. July–August.
ECOLOGY. Rocky places, open woodland, scrub, sea level to 660 m (and probably
higher).
DISTRIBUTION. Former Yugoslavia; Albania; Bulgaria; Turkey-in-Europe.

80d subsp. **dilatatum** *(Zahariadi)B.Mathew* comb. et stat. nov. Type: Crete, 'in pinetis Samariae', 400–800 m, *W.Greuter* 9419 (holotype G, isotype ATH[22268]!).

A.dilatatum Zahariadi in Ann. Mus. Goulandris 3:88(1977); Stearn in Ann. Mus. Goulandris 4:186(1978); Stearn in Fl. Europaea 5:67(1980).

ILLUSTRATIONS. Ann. Mus. Goulandris 3:89,f.20(1977).

Perianth segments white with a green median stripe, the inner three 5 mm long.

FLOWERING TIME. July–August.
ECOLOGY. Dry rocky places and open pine woods, 400–800 m.
DISTRIBUTION. Crete, recorded in Hania and Rethymnon provinces.

81. Allium amethystinum *Tausch* in Syll. Pl. Nov. Ratisb. (Konigl. Baier. Bot. Ges.) 2:256(1828); Bothmer in Bot. Not. 125:70(1972); Garbari & Corsi in Inform. Bot. Ital. 4:125(1972); Stearn in Ann. Mus. Goulandris 4:183(1978); Stearn in Fl. Europaea 5:67(1980); Kollmann in Davis, Fl. of Turkey 8:186(1984). Type: 'Dalmatia' (lectotype PRC!, see Bothmer in Bot. Not. 125:70(1972)).

A.segetum Jan ex Schultes & Schultes fil. in Roemer & Schultes, Syst. Veg. 7:1020(1830); Kollmann in Notes R.B.G. Edinb. 31:119(1971). Type: Sicily, *Jan* (not traced).
A.rollii Terrac. in Malpighia 3:289(1889). Type: Italy, 'in aridis, secus viam Aureliam circa Romam "dall'Acqua fredda alla Maglianella", et in montosis apud "Corneto"', June 1866, *Rolli* 10, (holotype RO!).
A.calyptratum Rolli in sched.
A.sphaerocephalum subsp. *rollii* (Terrac.)K.Richter, Pl. Eur. 1:200(1890).
A.stojanovii Kovatschev in Nov. Syst. Pl. Vasc. (Leningrad) 1968:42(1968). Type: Bulgaria, Primorsko, 22 July 1966, *I.Kovatschev* (holotype Plovdiv, isotype LE).
'*A.descendens*' auct. non L., e.g. Boiss., Fl. Or. 5:236(1882); Halacsy, Consp. Fl. Graecae 3:243(1904).

ILLUSTRATIONS. Notes R.B.G. Edinb. 31:120,f.1a;121,f.2a-e(1971) [as *A.segetum*]; Bot. Not. 125:64,f.D&E;73,f.11(1972); Fl. of Turkey 8:183,f.7,no.9(1984). Plate 13A-B.

Bulb subglobose, 1.5–2 cm diam.; outer tunics whitish, membranous; bulblets absent or few, ovoid, yellowish or pale brown, the surface (beneath the outer tunic) reticulate-patterned, carried above the bulb on a long stipe beneath the leaf sheaths. Stem (30–)50–120 cm, often purplish, smooth. Leaves (3–)4–5(–7), shorter than the inflorescence, linear, canaliculate, fistulose, often with a pale median stripe on the upper (adaxial) surface, 2–8 mm wide, with sparse teeth on the margins, sheathing the lower quarter to third of the stem, usually withered away at anthesis; sheaths smooth. Spathe 1-valved, up to 8 cm long, broadly ovate at the base, narrowing abruptly to a long beak, caducous. Umbel spherical, 2.5–6.5 cm diam. Pedicels smooth, unequal, 1–2 cm long at anthesis, the inner ones then soon becoming erect and elongating to up to 5 cm, forming the impression of a 'two-tiered' umbel; bracteoles present, forming a small, membranous silvery-white involucre around the base of the umbel. Perianth cylindrical-campanulate; segments purple or dull pinkish-purple, narrowly oblong, spathulate or oblong-elliptical, 3–5(–6) mm long, obtuse, truncate or emarginate, smooth, the outer slightly shorter and broader than the inner. Stamens with exserted anthers; filaments usually minutely ciliate-dentate, the outer ones lanceolate-subulate, the inner ones with the anther-bearing cusp slightly shorter to slightly longer than the expanded undivided basal part and about half to two thirds as long as the lateral cusps; lateral cusps long-exserted; anthers yellow or purple before dehiscence. Style exserted. Capsule subglobose, about 4–5 mm long; seeds 3–4.5 mm long, flattish, black. 2n=16.

FLOWERING TIME. May–June.

ECOLOGY. Rocky places, maquis, open *Pinus* woods, cultivated and abandoned fields, sea level–1600 m.

DISTRIBUTION. Italy; Sicily; former Yugoslavia; Albania; Bulgaria; Greece; Crete; E Aegean Is.; NW, W, SW Turkey.

NOTES. *A.amethystinum* is rather similar to *A.guttatum* and there is a discussion about the differences to be found under the latter species. Bothmer in Bot. Notiser 125:75(1972) draws attention to the fact that *A.amethystinum* has relatively large flat seeds whereas those of *A.guttatum* are smaller and angular-triquetrous; additionally, the bulblets of the former have a reticulate pattern on the surface while those of *A.guttatum* are nearly smooth.

82. Allium affine *Ledeb.*, Fl. Ross. 4:166(1852); Grossh., Fl. Kavk. ed.1,1:202 (1928), ed.2,2:123(1940); Vved. in Fl. URSS 4:239(1935), Engl. ed. 4:185(1968); Feinbr. in Pal. Journ. Bot., Jer. Ser. 3:11(1943); Mouterde, Nouv. Fl. Lib. et Syr. 1:267(1966); Wendelbo in Rechinger, Fl. Iranica 76:52(1971); Kollmann in Davis, Fl. of Turkey 8:188(1984); Wendelbo in Townsend & Guest, Fl. of Iraq 8:152(1985). Type: Caucasus, 'Iberia', *Wilhelms* (syntype LE, photo K!); 'prope Tiflis', *C.Koch* (syntype B, prob. destroyed); 'Somchetia', *Hohenacker* (syntype LE).

A.margaritaceum var. *affine* (Ledeb.)Regel, All. Monogr. 50(1875); Boiss., Fl. Or. 5:240(1882).

A.margaritaceum var. *scabrum* Regel, All. Monogr. 51(1875). Type: partly based on the *Wilhelms* specimen from 'Iberia', cited above as one of the syntypes of *A.affine*.

A.affine var. *scabrum* (Regel)Grossh., Fl. Kavk. ed.1,1:202(1928).

?*A.mishtschenkoanum* Grossh. in Grossh. & Schischkin, Pl. Or. Exs. no.4(1924). Type: Transcaucasia, Erivan, 19 July 1919, *Grossheim* (isotype K!).

ILLUSTRATIONS. Mouterde, Nouv. Fl. Lib. et Syr. 1, Atlas:t.84,f.1(1966); Fl. Iranica 76:t.5,f.70(1971); Fl. of Turkey 8:183,f.7,no.12(1984). Plate 13C.

Bulb ovoid, 1–2.5 cm diam.; outer tunics greyish, membranous, sometimes splitting lengthways; bulblets lacking or few, yellowish. Stem 30–80 cm, pale greyish-green, rather thick and increasing in diameter up to the umbel where it may be up to 5 mm diam. Leaves 3–5, sheathing the lower third to half of the stem, rarely to about two thirds, linear, semi-cylindrical, canaliculate, fistulose, 2–4 mm wide, the margin minutely scabrid; sheaths prominently ribbed, smooth or scabrid. Spathe 1-valved, with an ovate base, narrowed into a long slender beak up to 13 cm long, scabrid on the veins, caducous. Umbel spherical, dense, 2.5–4(–4.5) cm diam. Pedicels unequal, slender, 1–2(–2.5) cm long, smooth; bracteoles present, filiform, up to 2.5 cm long, silvery-white, very conspicuous or sometimes minute (see note below). Perianth cylindric-campanulate or ovoid-campanulate; segments whitish with a green median vein or green with narrow white margins, 3–4 mm long, smooth, the outer oblong or elliptic-ovate, obtuse or emarginate, the inner oblong-linear, slightly longer and narrower than the outer, obtuse or retuse. Stamens exserted; filaments ciliate at the base, the outer ones simple, narrowly triangular, the inner ones with the anther-bearing cusp about equal in length to the expanded undivided basal lamina and a third to half as long as the lateral cusps, lateral cusps well-exserted; anthers greenish-yellow or yellow. Style included or shortly exserted at anthesis. Capsule ellipsoid-subglobose, 3–4 mm long; seeds 2–3 mm long, black. 2n=16.

FLOWERING TIME. June–August.

ECOLOGY. Stony slopes, in dry scrub, steppe, oak-pine woods, 500–2000m.

DISTRIBUTION. S & E Transcaucasus; C-S & E Turkey; N & NW Iran; N Iraq; Syria; Lebanon.

NOTES. *A.affine* is undoubtedly related to *A.guttatum* but may be distinguished by having a very much longer beak to the spathe, and longer filiform bracteoles; the distribution of *A.affine* is somewhat to the east of that of *A.guttatum*, and the former is an Irano-Turanian species while the latter belongs primarily to the Mediterranean element.

Some specimens from Iraq and SE Turkey [eg. *Stevens* 29(K!); 41(K!)], and the Grossheim type specimen of *A.mishtschenkoanum*, lack the long filiform bracteoles which are such a noticeable feature of *A.affine*; further studies are required to ascertain whether or not these should be treated as a separate taxon. The plant described as *A.transcaucasicum* [in Russian] by Grossheim, Fl. Kavk. ed.2,2:123(1940), requires assessment; in the key to the species it was distinguished from *A.artvinense* on account of the stem being 'inflated' in the upper part (ie., as in *A.affine*), the pedicels subequal in length, and a slightly differently shaped umbel. The type locality is Lerik in the Talysh region of Transcaucasia, but I have not ascertained the whereabouts of the type specimen. However, in Tbilisi there is a specimen which has been annotated *A.transcaucasicum* by Grossheim in 1940 ['Nachitschevan, prop. Milach', 9 August 1940, *Karjagin* (TBI!)]; this does have an apically-swollen stem and nearly all of its features are in agreement with those of *A.affine* but, like the specimens mentioned above, it lacks the long thread-like bracteoles which are such a striking feature of the latter species.

83. Allium gorumsense *(Regel)Boiss.*, Fl. Or. 5:233(1882); Kollmann in Davis, Fl. of Turkey 8:189(1984). Type: Turkey, 'ad pagum Gorumse Ciliciae Kurdicae', 1280 m, 1859, *Kotschy* 174 (holotype G!).

A.margaritaceum var. *gorumsense* Regel, All. Monogr. 51(1875).

Bulb ovoid; outer tunics membranous; Stem about 60 cm. Leaves shorter than the inflorescence, linear, fistulose, many-veined, about 4.5 mm wide, glabrous, sheathing about half of the stem. Spathe caducous. Umbel subglobose, densely-flowered, 3–4 cm diam. Pedicels slender 1.2–1.8 cm long; bracteoles present. Perianth campanulate; segments about 4 mm long, white, ovate-oblong, acute or subacute, obscurely keeled, glossy. Stamens exserted; outer filaments simple, narrowly lanceolate, ciliate at the base, the inner ones glabrous, with the anther-bearing cusp about as long as the undivided basal part and slightly shorter than the more slender lateral cusps; lateral cusps exserted from the perianth; anther colour unknown. Style exserted. Capsule details unknown.

FLOWERING TIME. May.

ECOLOGY. The type specimen was collected at about 1280 m but no details of habitat were noted.

DISTRIBUTION. Turkey, recorded only in Adana province.

NOTES. The species is known only from the type collection so that it is impossible to be sure of its relationships at present.

84. Allium chamaespathum *Boiss.*, Diagn. Pl. Or. Nov. 1,7:113(1846); Regel, All. Monogr.:48(1875); Boiss., Fl. Or. 5:238(1882); Halàcsy, Consp. Fl. Graecae 3:244(1904); Hayek, Prodr. Fl. Penins. Balcan. 3:43(1932); Bothmer in Bot. Not. 125:68(1972); Stearn in Ann. Mus. Goulandris 4:187(1978); & in Fl. Europaea 5:68(1980). Syntypes: Greece, Attica, Mt Hymettus, *Spruner* (syntype G); Zakynthos Is., *Margot* (syntype ?G).

ILLUSTRATIONS. Bot. Not. 125:68,f.9(1972). Plate 13D.

Bulb ovoid, 1.8–2.5 cm diam.; outer tunics greyish, membranous; bulblets usually absent but bulb sometimes splitting into two after flowering. Stem (10–)25–30(–60) cm, smooth. Leaves 2–3, the upper one exceeding the inflorescence, linear, fistulose, canaliculate on the upper surface, 2–4 mm wide, smooth, sheathing the whole of the stem up to the umbel; sheaths smooth but conspicuously ribbed. Spathe 2.5–5 cm long, 1-valved, beaked, deciduous but often retained by the sheathing upper leaf. Umbel spherical, densely-flowered, 3–4.5 cm diam. Pedicels 1–1.5 cm long, subequal, smooth, without bracteoles. Perianth cylindrical; segments white or greenish, 4–4.5 mm long, narrowly oblong, smooth, the outer rounded at the apex, the inner truncate. Stamens slightly exserted; filaments minutely ciliate at the base, the outer ones simple, narrowly triangular, the inner ones with the anther-bearing cusp half as long as the undivided basal part and equalling or slightly longer than the lateral cusps; lateral cusps slightly exserted from the perianth; anthers yellow. Style well-exserted. Capsule subglobose, 4–5 mm long; seeds 3–4 mm long. 2n=16.

FLOWERING TIME. (July–)August–October.
ECOLOGY. Rocky (calcareous) mountain slopes, sea level to 2135 m
DISTRIBUTION. Greece, incl. Crete; Albania.
NOTES. A very distinct member of section *Allium*, instantly recognisable on account of the fact that the stem is entirely sheathed all the way up to the umbel, with the free apex of the upper leaf protruding from the umbel in a spathe-like manner.

85. Allium vineale *L.*, Sp. Pl. 1:299(1753); Sowerby & Smith, Engl. Bot. 28:t.1974(1809); Reichenb., Ic. Fl. Germ. 10:23,t.490(1848); Syme, Engl. Bot. 9:t.1534(1869); Regel, All. Monogr. 40(1875); Boiss., Fl. Or. 5:235(1882); Fiori & Paol., Ic. Fl. Ital. 79(1898); Coste, Fl. France 3:334(1906); Hegi, Ill. Fl. Mittel-Eur. 2:217(1909); Hayek, Prodr. Fl. Penins. Balcan. 3:44(1932); Vved. in Fl. URSS 4:237(1935), Engl. ed. 4:183(1968); Wilde-Duyfjes, Revis. Allium in Africa 29(1977); Stearn in Ann. Mus. Goulandris 4:182(1978); Stearn in Fl. Europaea 5:67(1980); Pastor & Valdes, Revis. Gen. Allium Penins. Iber. & Isl. Bal. 70(1983); Kollmann in Davis, Fl. of Turkey 8:185(1984). Type: 'Habitat in Germania'. For discussions about typification see Wilde-Duyfjes in Taxon 22:88(1973) and Pastor & Valdes, Revis. Gen. Allium Penins. Iber. & Isl. Bal. 72(1983).

A.arenarium L., Sp. Pl. 1:297(1753). Type: 'Habitat in Thuringia; in Falstria Scaniae'[LINN].
A.compactum Thuill., Fl. Paris, ed.2:167(1800). Type: not stated.
Porrum arenarium (L.)Reichenb., Fl. Germ. Excurs. 114(1830).
A.purshii G.Don, Monogr. Allium 10(1832). Type: 'Habitat in Canada'.
Getuonis vineale (L.)Rafin., Fl. Tellur 20(1836).
A.vineale var. *capsuliferum* Koch, Syn. Fl. Germ. ed.2:831(1844). Type: not stated.
A.vineale var. *compactum* (Thuill.)Cosson, Germ. & Wedd., Intr. Fl. Anal. Descr. Env. Paris 128(1842). Type: as for *A.compactum* Thuill.
A.affine Boiss. & Heldr. in Boiss., Diagn. Pl. Or. Nov. 2,4:114(1859), non

Ledeb.(1852). Type: Greece, Mt. Parnassus, 9–10 August 1852, *Heldreich* (lectotype G, see Wilde-Duyfjes, Revis. Allium in Africa 30(1976)).

A.vineale var. *purpureum* H.P.G.Koch in Vidensk. Medd. Naturh. Foren. Kjøbenhavn 1862:117(1863). Type: Denmark, Nykjøbing to Veggerløse, *H.P.G.Koch* (C).

A.nitens Sauze & Maillard, Cat. Fl. Deux-Sevres 51(1864). Type: France, Parthenay, Airvault (not traced).

A.kochii Lange, Haandbog Danske Fl. ed.3:254(1864). Type: as for *A.vineale* var. *purpureum* H.P.G.Koch.

?*A.vineale* var. *asperiflorum* Regel, All. Monogr. 41(1875). Type: Hungary, 'in campis arenosis territorii Bakos' (not traced).

A.vineale var. *virens* Boiss., Fl. Or. 5:236(1882). Type: as for *A.affine* Boiss. & Heldr.

A.vineale subsp. *compactum* (Thuill.)K.Richter, Pl. Eur. 1:198(1890). Type: as for *A.compactum* Thuill.

A.vineale subsp. *capsuliferum* (Koch)K.Richter, Pl. Eur. 1:199(1890). Type: as for *A.vineale* var. *capsuliferum* Koch.

A.vineale subsp. *affine* (Boiss. & Heldr.)K.Richter, Pl. Eur. 1:198(1890). Type: as for *A.affine* Boiss. & Heldr.

A.margaritaceum var. *compactum* Batt. & Trabut, Fl. Alger. Monocot. 61(1895). Type: Algeria, Le Corso, *Battandier* (holotype MPU).

A.margaritaceum var. *bulbiferum* Batt. & Trabut in Bull. Soc. Bot. France 51:353(1904). Type: Algeria, forêt d'Afir, *Battandier* (holotype P).

A.vineale var. *typicum* Aschers. & Graebn., Syn. Mitteleur. Fl. 110(1905).

A.assimile Halàcsy, Consp. Fl. Graecae 3:249(1904). Syntypes: Greece: 'Neuropolis, *Haussknecht*; Mt.Tymphrestos, *Samaritani*; Mt.Parnassus, *Heldreich*; Mt.Kyllene, *Orphanides.*'

A.vineale var. *nitens* (Sauze & Maillard)Coutinho, Fl. Port. 136(1913). Type: as for *A.nitens* Sauze & Maillard.

A.gaditanum var. *bulbiferum* Coutinho, Fl. Port. 128(1913). Type: Portugal, Fundao, July 1922, *L.Fernandes* LISU 65735 (lectotype LISU, see Pastor & Valdes, Revis. Gen. Allium Penins. Iber. & Isl. Bal. 72(1983)).

ILLUSTRATIONS. Sowerby & Smith, Engl. Bot. 28:t.1974(1809); Reichenb., Ic. Fl. Germ. 10:t.490,f.1075(1848); Hegi, Ill. Fl. Mittel-Eur. 2:217,f.336(1909); Ross-Craig, Draw. Brit. Pl. 29:t.19(1972); Wilde-Duyfjes, Revis. Allium in Africa 31,f.3(1976); Fl. of Turkey 8:183,f.7,no.9(1984). Plate 13E.

Bulb ovoid, 0.8–2 cm diam.; outer tunics greyish-brown, membranous, splitting longitudinally into strips; bulblets present, yellowish. Stem 30–120 cm, smooth or minutely scabrid on the ribs. Leaves (3–)4–5(–7), usually shorter than the inflorescence but sometimes overtopping it, subterete, canaliculate, fistulose, 1.5–5 mm wide, glabrous, sheathing the lower third to two thirds of the stem; sheaths smooth. Spathe 1-valved, 1–4.5 cm long, broadly ovate at the base, narrowing abruptly to a beak equal in length to, or slightly longer than, the broad base, caducous. Umbel spherical, ovoid or hemispherical, 2–5 cm diam., wholly floriferous, partly bulbilliferous or wholly bulbilliferous; bulbils very variable in shape and size, subglobose to fusiform, sessile or subsessile, usually shortly beaked. Pedicels smooth, unequal, 0.5–3 cm long; bracteoles usually forming a small, membranous silvery-white involucre around the base of the umbel, sometimes also a few minute linear bracteoles present. Perianth campanulate; segments pink, purple, purplish-red or greenish-white, 2–3.5(–5) mm long, smooth, the outer strongly concave, narrowly ovate, subacute or obtuse, the inner narrowly oblong or spathulate-obovate, rounded or obtuse. Stamens with partially to distinctly exserted anthers; filaments minutely ciliate-dentate at the base, the outer ones triangular-subulate, the inner ones with the anther-bearing cusp equalling or slightly shorter than the expanded undivided basal part and about half to two thirds as long as the lateral cusps; lateral cusps exserted; anthers yellow. Style exserted. Capsule ovoid, about 3–3.5 mm long; seeds apparently seldom developed. 2n=16,32,40,48.

FLOWERING TIME. June–August.

ECOLOGY. Meadows, marshes, riversides, waste places, cultivated fields, roadsides, sea level–2650 m.

DISTRIBUTION. Widespread in Europe and Turkey; Russia, (European part); Transcaucasia; Syria; Lebanon; NW.Africa; Canary Is.; naturalised in N.America.

NOTES. *A.vineale* is extremely variable and widespread on account of its ability to produce many bulbils which are readily distributed by the agricultural activities of man.

A.vineale commonly has bulbils present in the umbels, with or without flowers, and this feature means that it is fairly readily distinguished from the related species *A.guttatum* and *A.amethystinum*. The wholly bulbilliferous variants of *A.vineale* may be confused at first sight with the wholly bulbilliferous variants of *A.sphaerocephalon* but may be recognised by the fact that the 1-valved spathe of the former is early-deciduous whereas it is persistent and usually 2-valved in the latter. For this treatment a wide view of *A.vineale* has been taken; if, however, it is necessary to refer to floriferous and bulbilliferous variants, the following names are available:

Umbel bearing flowers only ....var. **capsuliferum** Koch
Umbel bearing flowers and bulbils ....var. **vineale**
Umbel bearing bulbils only ....var. **compactum** (Thuill.)Coss.

86. Allium subvineale *Wendelbo* in Rechinger, Fl. Iranica 76:53(1971); Matin, Contr. a l'étude de la Fam. Alliaceae en Iran 36(1976). Type: Iran, Gilan prov., Yehlah, Ispili, August 1936, *Lindsay* 1162 (holotype BM, isotype K!).

Bulb subglobose, about 1 cm diam.; outer tunics greyish-brown, membranous, extended into a long neck at the apex; bulblets not apparent. Stem 45–88 cm, smooth, prominently veined in the dried state. Leaves 2–3, shorter than the inflorescence, sub-terete, fistulose, canaliculate, 1–2.5 mm wide, ribbed, smooth, sheathing the stem for three quarters of its length; sheaths smooth. Spathe not seen, caducous. Umbel spherical, 2–4 cm diam., either many-flowered and dense or with many bulbils and few-flowered. Pedicels very unequal, up to 1.6 cm long; bracteoles present, ovate to linear, about 2 mm long. Perianth tubular-campanulate; segments white or pale pinkish-lilac, 3–3.5 mm long, elliptic-oblong, subacute, smooth. Stamens with the anthers slightly exserted; filaments glabrous, the outer ones simple, triangular-subulate, the inner ones with the antherbearing cusp slightly shorter than the expanded undivided basal part and about half as long as the lateral cusps; lateral cusps included; anthers violet. Style exserted. Capsule subglobose, about 3 mm long; seeds about 2 mm long, black.

FLOWERING TIME. August.

ECOLOGY. Meadows, damp grassland, boggy places.

DISTRIBUTION. Iran, recorded in the provinces of Bakhtiari, Gilan, Hamadan, Kordestan, Mazandaran, Semnan, Zanjan (after Matine l.c.).

NOTES. *A.subvineale* is not a well-known species and requires further study in the field. Kollmann in Davis, Fl. of Turkey 8:185(1984) considered it conspecific with *A.vineale*.

87. Allium stearnianum *Koyuncu, N.Özhatay* & *Kollmann* in Notes R.B.G. Edinb. 41:226(1983).

Bulb ovoid or globose, 1–1.5(–2.5) cm diam.; outer tunics brown, or brownish-grey, membranous or subcoriaceous, sometimes splitting into parallel fibres near the apex; bulblets apparently absent. Stem 20–35(–40) cm, smooth, conspicuously glaucous. Leaves 1–3, much shorter than the inflorescence, linear, fistulose, flattened or canaliculate on the upper (adaxial) side, 2–3(–4) mm wide, smooth or scabrid-papillose on the margins, sheathing the lower fifth to third of the stem; sheaths smooth but conspicuously ribbed or winged. Spathe 1-valved, ovate at the base, narrowing abruptly to a short beak, caducous. Umbel spherical, 1–2.5 cm diam., very dense. Pedicels smooth, unequal, up to 1 cm long; bracteoles present, silvery-white, laciniate. Perianth campanulate or ovoid-campanulate; segments deep purple in the upper half, paler below, with a green or purple median vein, 2.5–4 mm long, oblong or ovate-oblong, keeled, obtuse, verrucose-scabrid, mainly on the keel. Stamens with long-exserted anthers; filaments glabrous or minutely ciliate-dentate, the outer ones subulate, the inner ones with the anther-bearing cusp equalling or slightly longer than the expanded undivided basal part and about two thirds as long as or almost as long as the lateral cusps; lateral cusps exserted; anthers purple before dehiscence. Style exserted. Capsule ovoid, 2–3 mm long.

NOTES. *A.stearnianum* may be recognised by its small very tight umbels of deep purple flowers which have verrucose perianth segments.

Two subspecies are recognised:
1. Stem zig-zag in the lower third; leaves smooth; perianth segments 2.5–3 mm long; ovary smooth ....subsp. **stearnianum**
   Stem straight; leaves with scabrid margins; perianth segments 3–4 mm long; ovary papillose ....subsp. **vanense**

87a. subsp. **stearnianum.** Type: Turkey, Sivas/Tokat, Çamlibel Dağ, 5 July 1980, *Koyuncu et al.* AEF 7456 (holotype AEF!).

ILLUSTRATIONS. Notes R.B.G. Edinb. 41,2:260,f.12B(1983); Fl. of Turkey 8:165,f.6,no.11(1984). Plate 14A-B.

Stem flexuose in the lower half. Leaves with smooth margins. Perianth segments 2.5–3 mm long. Ovary smooth. 2n=16.

FLOWERING TIME. July.
ECOLOGY. Rocky places, 1600–1750 m.
DISTRIBUTION. C.Turkey, recorded in Sivas, Tokat and Gümüşane provinces.

87b. subsp. **vanense** *Kollmann & Koyuncu* in Notes R.B.G. Edinb. 41,2:267(1983); Kollmann in Davis, Fl. of Turkey 8:171(1984). Type: Turkey, Van province, Gürpinar to Başkale, Çuh Pass, 2750 m, *T.Baytop* ISTE 45342 (holotype ISTE!).

ILLUSTRATIONS. Notes R.B.G. Edinb. 41,2:261,f.13(1983); Fl. of Turkey 8:165,f.6,no.12(1984).

Stem more or less straight. Leaves with scabrid margins. Perianth segments 3–4 mm long. Ovary densely papillose.

FLOWERING TIME. July–August.
ECOLOGY. Dry rocky hillsides, 2500–3000 m.
DISTRIBUTION. SE. Turkey, recorded in Van and Hakkâri provinces.

88. Allium hierochuntinum *Boiss*., Fl. Or. 5:244(1882); Post, Fl. Syria, Palestine & Sinai ed.2,2:637(1933); Feinbr. in Pal. Journ. Bot., Jer. Ser. 3:20(1943); Mouterde, Nouv. Fl. Lib. et Syr. 1:271(1966); Kollmann in Rotem 15:75(1985); Kollmann in Feinbr., Fl. Palaestina 4:93(1986). Types: Israel, 'in faucibus calidis Palestinae inter Hierosolymam et Jericho', *Boissier* (syntype G); 'infra monasterium St. Saba', *Kotschy* 436 (lectotype G, isolecto-type K!,P!); Syria, 'ad Snou Fadel montis Gebel Belas deserti prope Palmyram', *Blanche* (syntype G).

A.ascalonicum L., Fl. Palaestina:17(1756). Type: Israel, probably between Jerusalem and Jericho, *Hasselquist* (LINN) [See Stearn, Bull. Brit. Mus., Bot. 2:181(1960)].

ILLUSTRATIONS. Rotem 15:72(1985); Fl. Palaestina 4(Plates):127(1986). Plate 14C.

Bulb oblong-ovoid, 0.8–1.5 cm diam.; outer tunics brown, conspicuously reticulate-fibrous; bulblets, if produced, ovoid-elongate, formed around the parent bulb or sometimes just below on a 'dropper'. Stem 12–15(–35) cm, smooth, rather slender and wiry. Leaves 2–4, usually shorter than the inflorescence but sometimes equalling it, filiform-terete, fistulose, 0.5–1.5 mm wide, smooth, sheathing the lower quarter to a third of the stem; sheaths smooth. Spathe 2-valved, the valves ovate, acuminate or shortly mucronate, membranous, often purplish-tinged, persistent. Umbel spherical, (1–)1.5–2.5(–3.5) cm diam., dense. Pedicels 4–10 mm long, smooth; bracteoles present, white, membranous. Perianth narrowly campanulate; segments violet blue, 6–8 mm long, ovate-lanceolate, tapering gradually to an acute or subacute apex, the outer ones papillose on the outer surface and scabrid on the keel and margins, the inner smooth or papillose. Stamens with the anthers included; filaments very slightly scabrid-ciliate at base, the outer ones simple, rather broad and triangular in the lower half, the inner ones with the anther-bearing cusp about a third as long as the widely expanded undivided basal part and about half as long as the lateral cusps; lateral cusps included within the perianth; anthers purple before dehiscence. Style included. Capsule globose-ovoid, 2.5–3.5 mm long; seeds 2.5 mm long. 2n=16.

FLOWERING TIME. March–May.
ECOLOGY. Steppe, desert, -200 to +750 m.
DISTRIBUTION. Israel, Jordan, Syria.
NOTES. This is very closely related to *A.scabriflorum* which has a more northern distribution in Turkey, but the latter has noticeably shorter perianth segments and more coarsely reticulate bulb tunics.

A.ascalonicum L.(1756) is clearly the earliest name for this species. However, Linnaeus, Species Plantarum ed.2:429(1762), altered his concept of the species and treated his type (*Hasselquist*, LINN) as representing the flowering stage of one of the old 'bunch-forming' cultivars of the Onion, *A.cepa*, one clone of which has become known as the Shallot. In this concept, *A.ascalonicum* L.(1762) has been used more or less continuously since that time as the correct name for the Shallot and its related forms, although the epithet and type actually refer to the quite different taxon which was later described as *A.hierochuntinum* by Boissier(1882). Article 69 of the International Rules of Nomenclature states that 'a name may be ruled as rejected if it has been widely and persistently used for a taxon or taxa not including the type.' Stearn(1960) presented the view that on these grounds *A.ascalonicum* L. should be rejected. Wilde-Duyfjes, A Revision of the genus Allium in Africa :54(1976), took the opposing view, that *A.ascalonicum* L. was acceptable, and that it represented a very widely distributed taxon, including *A.scabriflorum* Boiss. from Turkey, *A.barthianum* Aschers. & Schweinf. from Libya and Egypt, and *A.artemisietorum* Eig & Feinbr. from Israel, as well as the species described above, *A.hierochuntinum* Boiss. Here, these taxa are treated as related, but distinct, and *A.ascalonicum* L. (1756) is rejected.

89. Allium scabriflorum *Boiss.*, Diagn. Pl. Or. Nov. 1,5:60(1844); Boiss., Fl. Or. 5:242(1882); Kollmann in Davis, Fl. of Turkey 8:192(1984). Type: Turkey, 'in Cappadocia Orientali', *Aucher* 2196 (holotype G,isotype P!).

ILLUSTRATIONS. Fl. of Turkey 8:183,f.7,no.18(1984). Plate 14D-E.

Bulb ovoid, 0.7–1(–1.5) cm diam.; outer tunics golden brown, strongly reticulate-fibrous; bulblets absent, or the parent bulb splitting into (usually two) replacement bulbs more or less the same size as the original and persisting within the same tunic, eventually forming clumps. Stem 10–25(–30) cm, smooth, rather slender and wiry. Leaves 2–3, usually shorter than the inflorescence but sometimes equalling it, filiform-terete, fistulose, shallowly canaliculate, 1–1.5 mm wide, smooth or slightly scabrid-papillose on the margin, sheathing the lower quarter to a third of the stem; sheaths smooth. Spathe ovate, acuminate or shortly mucronate, whitish, membranous, persistent, entire or splitting into 2 or more lobes. Umbel fastigiate-hemispherical to spherical, 0.8–2.5(–3) cm diam., dense. Pedicels (2–)3–10 mm long, smooth; bracteoles present (sometimes absent on weak, few-flowered specimens), white, membranous. Perianth campanulate; segments usually violet blue with a greenish median vein or sometimes deep purple and tending to dry bluish, or occasionally whitish with a green vein, 4–5 mm long, ovate-lanceolate, tapering gradually to an acute or rarely subacute apex, the outer ones papillose on the outer surface and scabrid or conspicuously crested on the very prominent keel, the inner smooth to papillose or crested. Stamens with anthers partially to fully exserted at anthesis; filaments very slightly scabrid-ciliate at base, the outer ones simple, rather broad and triangular in the lower half, the inner ones with the anther-bearing cusp about a third as long as the widely expanded undivided basal part and about two thirds as long as the lateral cusps; lateral cusps slightly to well-exserted; anthers usually purple before dehiscence but apparently occasionally yellow. Style included to slightly exserted. Capsule ovoid, 3 mm long; seeds 2.5 mm long. 2n=16.

FLOWERING TIME. June–July.
ECOLOGY. Saline steppe, dryish sandy places, 700–1700 m.
DISTRIBUTION. C.& S.Turkey (recorded in Adana, Afyon, Amasya, Ankara, Burdur, Eskişehir, İçel (Mersin), Konya, Malatya, and Niğde provinces).
NOTES. This closely resembles *A.hierochuntinum*, which occurs farther south in Jordan, Israel and Syria, but has shorter perianth segments and more coarsely reticulate bulb tunics.

90. Allium armerioides *Boiss.*, Diagn. Pl. Or. Nov. 1,7:116(1846); Boiss., Fl. Or. 5:244(1882); Kollmann in Davis, Fl. of Turkey 8:192(1984). Type: Turkey, Mardin prov., 'in Assyria inter Diarbekir et Mardin', *Kotschy* 1843:286 (holotype W, isotype G, photoK!).

Bulb ovoid, about 1 cm diam.; outer tunics brown, finely reticulate-fibrous, produced into a distinct neck. Stem 10–14 cm. Leaves approximately equalling the stem, filiform, fistulose, 1–2 mm wide. Spathe 2-valved, the valves ovate, acute, membranous, equalling the umbel, persistent. Umbel spherical, 1–1.5 cm diam. few-flowered. Pedicels up to 5 mm long. Perianth campanulate; segments white with a green midvein, 5–6 mm long, papillose on the outside and aculeolate on the keel. Stamens shortly exserted, the outer ones simple, inner ones with the anther-bearing cusp about a third as long as the undivided basal part and somewhat shorter than the lateral cusps; lateral cusps slightly exserted from the perianth; anther colour unknown. Capsule details unknown.

FLOWERING TIME. probably June.

ECOLOGY. Unknown.

DISTRIBUTION. S.Turkey (recorded in Mardin province).

NOTES. Known only from the type collection; it appears to be somewhat similar to *A.scabriflorum* but has a few-flowered umbel of larger white flowers.

91. Allium artemisietorum *Eig & Feinbr.* in Pal. Journ. Bot., Jer. Ser. 3:18(1943); Mouterde, Nouv. Fl. Lib. et Syr. 1:271(1966); Kollmann in Rotem 15:75(1985); Kollmann in Feinbr., Fl. Palaestina 4:94(1986). Type: Israel, N.Negev, about 10 km E. of Beer-Sheba, Eocene hills, *Artemisietum herbae-albae*, 29 May 1942, *N.Feinbrun* (holotype HUJ!).

ILLUSTRATIONS. Pal. Journ. Bot., Jer. Ser. 3:15,f.8 & 19,f.30(1943); Rotem 15:74(1985); Fl. Palaestina 4(Plates):128(1986). Plate 15A.

Bulb oblong-ovoid, (1–)1.5–2.5(–3) cm diam.; outer tunics brown, reticulate-fibrous extended into a neck at the apex; bulblets few. Stem 15–30(–40) cm, smooth, rather flexuose. Leaves 2–3, usually shorter than the inflorescence, filiform-terete, fistulose, 1–2 mm wide, smooth, sheathing the lower quarter of the stem, usually withering before anthesis; sheaths smooth. Spathe 2-valved, the valves ovate, shortly mucronate, membranous, persistent. Umbel somewhat fastigiate-spherical,(1–)1.5–3(–4) cm diam., dense. Pedicels 3–12(–20) mm long, smooth; bracteoles present, white, membranous. Perianth campanulate; segments white with a reddish or green mid-vein, 3.5–5 mm long, ovate-oblong, subacute, scabrid on the keel and sometimes all over the outer surface, the inner smooth or papillose. Stamens with the anthers exserted; filaments glabrous, the outer ones simple, rather broad and triangular in the lower half, the inner ones with the anther-bearing cusp about a third as long as the widely expanded undivided basal part and about equal to the lateral cusps (i.e. slightly shorter to slightly longer); lateral cusps exserted from the perianth; anthers purple before dehiscence. Style exserted. Capsule ovoid, 3 mm long; seeds 2.5 mm long. 2n=16.

FLOWERING TIME. March–May.

ECOLOGY. Steppe and rocky places, desert, c. 360 m.

DISTRIBUTION. Israel, Jordan, Saudi Arabia, ?Egypt.

NOTES. This is related to *A.hierochuntinum* and *A.scabriflorum* but has differently coloured flowers with exserted anthers, the filaments are glabrous and the anther-bearing cusp of the inner filaments is almost equal to the lateral cusps.

92. Allium vuralii *Kit Tan* in Pl. Syst. Evol. 155:102(1987); Kit Tan in Davis, Fl. of Turkey 10:222(1988). Type: Turkey, Konya, Yavşantuzlasi, 16 October 1982, *Kit-Tan & Vural* 1557a (holotype E!).

ILLUSTRATIONS. Pl. Syst. Evol. 155:102,t.2E(1987).

Bulb oblong-ovoid, 1.5–2 cm diam.; outer tunics reticulate-fibrous. Stem 30 cm. Leaves 2, as long as or longer than the inflorescences, filiform, fistulose, 1.5–2 mm wide, glabrous. Spathe 2-valved, membranous, the valves more or less equal, ovate, 0.5–1 cm long. Umbel spherical, densely-flowered, about 2.5 cm diam. Pedicels about 1 cm long. Perianth campanulate-urceolate; segments pale bluish violet with a green mid-vein, about 5 mm long, elliptic-ovate to obovate, acute to subacute, the outer ones papillose on the keel. Stamens with included or subexserted anthers; outer filaments simple, the inner ones with the anther-bearing cusp more or less equal to the lateral cusps; lateral cusps equalling the perianth or slightly exserted; anthers creamy-white. 2n=16.

FLOWERING TIME. July

ECOLOGY. Saline flats, about 850 m.

DISTRIBUTION. Turkey (recorded in the province of Konya).

NOTES. This is known only from the type collection; it is said to be related to *A.scabriflorum* Boiss. but differs in the shape and proportions of the inner filaments.

93. Allium barthianum *Aschers. & Schweinf.* in Bull. Herb. Boiss. 1,9:670(1893); Pamp., Prodr. Fl. Cir. 152(1931); Maire, Fl. de L'Afr. du Nord 5:267(1958); El-Gadi in Jafri & El-Gadi, Fl. of Libya 33:15(1977). Type: Libya, Cyrenaica, Badia, *Schweinfurth* 114 (lectotype G).

ILLUSTRATIONS. Maire, Fl. de l'Afr. du Nord 5:268,f.898(1958); Fl. of Libya 33:18,f.9(1977).

Bulb oblong-ovoid, 1–1.5 cm diam.; outer tunics yellow-brown, reticulate-fibrous, produced into a long neck at the apex; bulblets few. Stem about 15 cm, smooth. Leaves 2–3, slightly shorter than or equalling the inflorescence, filiform-terete, fistulose, 0.5–1 mm wide, smooth, sheathing the lower half of the stem; sheaths smooth. Spathe 2-valved, the valves ovate, shortly mucronate, scarious, white veined brownish-red, persistent. Umbel spherical, 1.5–3 cm diam., dense. Pedicels unequal, up to 10 mm long, smooth; bracteoles present, small, silvery-white. Perianth cylindric-campanulate; segments white with a green or brown median vein, about 5 mm long, oblong-lanceolate, acute, the outer ones scabrid on the keel. Stamens with the anthers included; filaments glabrous, the outer ones simple, triangular-subulate, the inner ones with the anther-bearing cusp about ?a third as long as the widely expanded undivided basal part and slightly longer than the lateral cusps; lateral cusps included within the perianth; anthers purple before dehiscence. Style included. Capsule not known.

FLOWERING TIME. March–April.

ECOLOGY. Dry rocky and gravelly places at low altitudes.

DISTRIBUTION. Libya, mainly Cyrenaica, but also recorded in Tripolitania.

NOTES. This is undoubtedly closely related to *A.hierochuntinum* from Israel but differs in that the median cusp of the inner filaments is longer than the lateral cusps; also the flower colour is white, not violet-blue. However, the species does require further study in the field to determine its variability and relationships with other species in this 'reticulate tunic/ fistulose leaf' group. Wilde-Duyfjes (1976) regards *A.barthianum* as a synonym of *A.ascalonicum*; comments upon this nomenclatural matter may be found in the notes under species no.88, *A.hierochuntinum*.

94. Allium sannineum *Gombault* in Bull. Soc. Bot. Fr. 84:470(1937); Mouterde, Nouv. Fl. Lib. et Syr. 1:271(1966); Kollmann & Shmida in Israel Journ. Bot. 26:135(1977); Kollmann in Rotem 15:75(1985). Type: Lebanon, Jebel Sannine, 1890 m, 20 July 1930, *Gombault* 3941 (holotype P!).

ILLUSTRATIONS. Rotem 15:73(1985).

Bulb ovoid, 0.7–1.5 cm diam.; outer tunics brown, reticulate-fibrous; bulblets present (according to Mouterde, op. cit.). Stem 10–20 cm, somewhat flexuose, with prominent veins, at least on drying. Leaves usually 3, equal to or longer than the inflorescence, terete, fistulose, 0.5–1.5 mm wide, sheathing a half to two thirds of the stem, usually still present at anthesis. Spathe 2-valved, the valves ovate, shortly acuminate, membranous, persistent, shorter than the umbel. Umbel spherical or subspherical, 1–1.5(–2) cm diam., fairly dense. Pedicels up to 5 mm long; bracteoles conspicuous, silvery-white, linear-lanceolate, exceeding the pedicels. Perianth campanulate; segments bluish-violet with a darker mid-vein, 4.5–5 mm long, the outer ones lanceolate, scabrid on the keel, subacute, the inner ones elliptic, obtuse. Stamens with the anthers exserted; outer filaments simple, triangular, the inner ones with the anther-bearing cusp about two thirds as long as the widely expanded undivided basal part and subequal in length to the lateral cusps; lateral cusps exserted from the perianth; anther colour brownish-yellow (in dried state). Style well-exserted. Capsule not seen. 2n=16.

FLOWERING TIME. June–August.

ECOLOGY. In alpine meadows and rocky (limestone) slopes, 1800–2800 m.

DISTRIBUTION. Lebanon, N.Israel(Mt. Hermon).

NOTES. Although somewhat similar to *A.scabriflorum* in outward appearance, the combination of reticulate bulb tunics, dwarf stature with leaves as long as the scape, and blue-violet flowers with exserted stamens, make this a fairly distinctive species; the high mountain habitat is also somewhat unusual.

95. Allium hamrinense *Hand.-Mazz.* in Ann. Naturh. Mus. Wien 28:15(1914); Zohary in Dep. Agr. Iraq Bull. 31:36(1950); Rawi in Dep. Agr. Iraq Tech. Bull. 14:185(1964); Wendelbo in Rechinger, Fl. Iranica 76:50(1971); Wendelbo in Townsend & Guest, Fl. of Iraq 8:148(1985). Type: Iraq, Upper Jazira, Jabal Makhul, S. of Shargat, 200–250 m, 10 May 1910, *Handel-Mazzetti* 1066 (holotype WU!).

ILLUSTRATIONS. Ann. Naturh. Mus. Wien 28:15,f.1(1914); Fl. Iranica 76:t.5,f.67(1971); Fl. of Iraq 8:149,pl.41(1985).

Bulb oblong, 1–1.5 cm diam.; outer tunics pale brown, reticulate-fibrous, produced into a neck up to 4 cm long; bulblets not apparent. Stem 34–70 cm, smooth. Leaves much shorter than the infloresence, withering away at or before flowering time, 2–3, cylindrical, fistulose, 1.5–3.5 mm wide, glabrous, sheathing the lower fifth to third of the stem. Spathe with 2–3 valves, the valves ovate, shortly acuminate, persistent. Umbel subglobose, 2–4 cm diam., dense. Pedicels 0.6–2.2 cm long at anthesis; bracteoles present, white. Perianth campanulate-urceolate; segments purplish-pink with a darker purple or green median vein, 3.5–4 mm long, elliptic, obtuse or retuse, scabrid on the exterior; margins papillose. Stamens with anthers at least partially exserted at anthesis; filaments glabrous, the outer ones simple, triangular, the inner ones with the anther-bearing cusp a third to half as long as the very widely expanded basal lamina and subequal to, or up to twice as long as, the lateral cusps; lateral cusps slightly shorter than or equalling the perianth; anthers violet. Style 3 mm long. Capsule subglobose, about 4 mm long; seeds black, 2–2.5 mm long.

FLOWERING TIME. May–June

ECOLOGY. Gravelly and sandy soils, often on calcareous formations, 200–750 m.

DISTRIBUTION. Iraq, mainly western subdesert regions.

NOTES. Related to the next species, *A.deserti-syriaci*, and apparently sharing a similar distribution and habitat; further notes can be found under that species. The type specimen has included anthers but was apparently collected at an early state, just prior to anthesis. It has more densely-scabrid perianth segments than the rest of the Iraqi material seen which is all from much farther to the south-west of the type location in the Western Desert region. Further collections in the type locality are therefore desirable.

96. Allium deserti-syriaci *Feinbr.* in Pal. Journ. Bot., Jer. Ser. 3:17(1943); Mouterde, Nouv. Fl. Lib. et Syr. 1:270(1966); Wendelbo in Townsend & Guest, Fl. of Iraq 8:151(1985). Type: Iraq, Jabal Jidra (Judram), between Damascus and Baghdad, 380km W. of Baghdad, 1 April 1933, *Eig & Zohary* (holotype HUJ!).

ILLUSTRATIONS. Pal. Journ. Bot., Jer. Ser. 3:19,f.29(1943); Nouv. Fl. Lib. et Syr. 1:t.84,n.6(1966).

Bulb ovoid, 1.5–2.5 cm diam.; outer tunics reticulate-fibrous, extended into a neck up to 5 cm long; bulblets present, few (according to Feinbrun, op. cit.). Stem 30–35 cm, flexuose. Leaves about 3, slightly shorter to slightly longer than the inflorescence, semi-terete, fistulose, 1.5–2.5 mm wide, smooth, sheathing the lower half of the stem, striate. Spathe broadly ovate narrowed to a beak 0.8–1.5 mm long, membranous, purplish with green veins, persistent. Umbel spherical or subspherical, 2.5–4.5 cm diam., fairly lax. Pedicels about 1–2 cm long; bracteoles present small, membranous. Perianth campanulate; segments white, 3.5–4 mm long, elliptic–ovate, retuse, smooth. Stamens with the anthers exserted; filaments glabrous, the outer ones simple, triangular-subulate, the inner ones with the anther-bearing cusp about as long as the expanded undivided basal part and equal in length to or slightly exceeding the lateral cusps; lateral cusps very shortly exserted from the perianth; anthers yellow. Style exserted. Capsule about 3–4 mm long.

FLOWERING TIME. April.
ECOLOGY. In desert situations in sandy or gravelly soils, 280–660 m.
DISTRIBUTION. Iraq, in the Western Desert region.
NOTES. A poorly-known species, having been collected only once since the type material. It is apparently somewhat related to *A.hamrinense* but differs in having white flowers with smooth perianth segments (i.e., not scabrid on exterior); furthermore, the inner filaments in *A.hamrinense* have the basal lamina much longer than the three cusps, whereas in *A.deserti-syriaci* it and the three cusps are all about equal in length.

97. Allium anatolicum *N.Özhatay & B.Mathew* in Kew Bull. 51(1): in press. Type: Turkey, Antalya province, Çobanisa to Korkuteli, 3 km from Çobanisa, 1050 m, 1 June 1988, *N.& E.Özhatay* ISTE 58856 (ISTE!).

ILLUSTRATIONS. Kew Bull. 51(1): in press.

Bulb ovoid, 1–1.5 cm diam., outer tunics brown, strongly reticulate-fibrous, produced at the apex into a short neck up to about 5 cm in length; bulblets few, golden-brown, elongate, acuminate. Stem about 30 cm. Leaves 3–4, as long as or slightly shorter than the stem, subterete, slightly channelled-flattened above, fistulose, 1–1.5 mm wide, prominently ribbed, scabrid on the margins and veins, sheathing the stem for a third of its length. Spathe 2-valved, the valves broadly ovate abruptly narrowed to a 4–5 mm long beak, membranous, persistent. Umbel dense, subspherical, about 3 cm diam. Pedicels 9–12 mm long, smooth; bracteoles present, about 2–3 mm long, membranous, white. Perianth campanulate or somewhat urceolate at anthesis; segments apparently white (in dried state), 3.5–4.5 mm long, ca. 1.5 mm wide, lanceolate, obtuse or emarginate, strongly keeled, the outer with sparse, prominent, branched, scabrid outgrowths on the keel, the inner smooth or sparsely scabrid. Stamens with anthers well-exserted; outer filaments simple, narrowly triangular, 5 mm long, smooth; inner filaments with the median cusp half to a third as long as the expanded basal lamina and slightly shorter than or equalling the lateral cusps, all three cusps well-exserted from the perianth at anthesis; anthers yellow. Style 3.5–4 mm long. Capsule and seeds not seen.

FLOWERING TIME. May–June
ECOLOGY. Rocky ridges, 1050 m.
DISTRIBUTION. SW Turkey (recorded in Antalya province).
NOTES. This is related to *A.robertianum* but is a much shorter plant with a 2-valved persistent spathe, the perianth segments have prominent branched outgrowths on the keel and the anthers are yellow.

98. Allium sinaiticum *Boiss.*, Diagn. Pl. Or. Nov. 1,13:31(1854); Fl. Or. 5:244(1882); Post, Fl. Syria, Palestine & Sinai ed.2,2:637(1933); Täckholm & Drar, Fl. Egypt 3:67(1954); Kollmann in Rotem 15:78(1985); Kollmann in Feinbr., Fl. Palaestina 4:94(1986). Type: Egypt, 'in locis elatis deserti Sinaitici inter conventum et jugum Tih', April 1846, *Boissier* (holotype G).

ILLUSTRATIONS. Rotem 15:76(1985); Fl. Palaestina 4(Plates):129(1986) [these illustrations are identical and show the anthers rather more prominently exserted than is actually the case]; Collenette, Ill. Guide to the Flowers of Saudi Arabia: 337(1985).

Bulb ovoid, 2–2.5 cm diam.; outer tunics brown, reticulate-fibrous; bulblets absent (?always). Stem 6–10(–15) cm, smooth, rather thick. Leaves 2(–3), longer than the inflorescence, terete, fistulose, 2–2.5 mm wide, smooth, sheathing the lower third to two thirds of the stem, usually still present at anthesis; sheaths smooth. Spathe 2-valved, the valves ovate, shortly mucronate, membranous and transparent with conspicuous veins, persistent. Umbel spherical or hemispherical, 2–4 cm diam., fairly dense. Pedicels 5–14 mm long, smooth; bracteoles present, white, membranous. Perianth campanulate; segments white with a reddish or green mid-vein, 6–7 mm long, lanceolate, scabrid-papillose all over the outer surface, the outer ones subacute, the inner ones obtuse or subacute and shortly mucronate. Stamens with the anthers included or partly exserted; filaments ciliate in the lower half, the outer ones simple, very broad and triangular, the inner ones with the anther-bearing cusp about a quarter as long as the very widely expanded ovate undivided basal part and two thirds to almost as long as the lateral cusps; lateral cusps included or subexserted from the perianth; anthers purple before dehiscence. Style included. Capsule ovoid, 4 mm long; seeds 3 mm long.

FLOWERING TIME. March–April.
ECOLOGY. In sand, semi-desert, c. 200–1200 m.
DISTRIBUTION. S Israel, SW Jordan(Edom), NW Saudi Arabia (Hijaz Prov.), NE Egypt (El Tiy, Sinai).
NOTES. This is a rather distinct species occurring in sandy places, often with the bulbs deeply seated and much of the stem buried.

99. Allium hedgei *Wendelbo* in Acta Horti Gotob. 28:20(1966); Wendelbo in Rechinger, Fl. Iranica 76:46(1971). Type: Afghanistan, Mazar-i Sharif, Takht-i Rostam prope Samangan, 1200 m, *Hedge & Wendelbo* 3990 (holotype BG!).

ILLUSTRATIONS. Acta Horti Gotob. 28:f.3(1966); Fl. Iranica 76:t.5,f.60 & t.16,f.1(1971).

Bulb ovoid, 1.5–2 cm diam., often several clustered together; outer tunics dark brown, strongly reticulate-fibrous, produced into a neck up to 2 cm long; bulblets pale brown, about 1 cm long, cymbiform. Stem 10–25 cm, sometimes 2 from each bulb, somewhat flexuose just above ground level. Leaves 1–3, equalling or slightly shorter than the inflorescences, linear, fistulose, canaliculate, ribbed, about 0.3–0.6 mm wide, minutely papillose on the margins, sheathing the lower fifth to third of the stem; sheaths smooth. Spathe with 2 reflexed, shortly caudate valves, up to 1.2 cm long. Umbel hemispherical, many-flowered, about 2.5–3 cm diam. Pedicels 1–2.3 cm long, smooth, thickened at the apex; bracteoles present. Perianth urceolate, umbilicate; segments purple-pink with a darker median vein, pruinose on the outside, 4.5–5 mm long, elliptic-ovate or ovate, obtuse or subacute, cucullate, smooth, the outer slightly shorter than the inner. Stamens with included anthers; filaments purple, smooth, the outer ones with two obtuse teeth at the base, subulate above, the inner ones with the anther-bearing cusp much longer than the undivided basal part and slightly longer to about twice as long as the lateral cusps; lateral cusps included within the perianth; anthers violet before dehiscence. Style purple, not exserted. Capsule ellipsoid, c. 3 mm long; seeds black, 2.5 mm long. 2n=16.

FLOWERING TIME. April–June
ECOLOGY. Stony slopes, 800–1200 m
DISTRIBUTION. N. & NE Afghanistan, in the regions of Samangan, Baghlan, Mazar-i Sharif and Kataghan.
NOTES. This is very distinct in having the inner filaments divided into three just above the base, and the outer filaments provided with small blunt teeth at the base.

100. Allium borszczowii *Regel* in Acta Horti Petrop. 3,2:74(1875); Wendelbo in Rechinger, Fl. Iranica 76:46(1971). Types: C.Asia, Syr Darya valley, *Borszczow* (syntype LE); near Mt. Karak, *O.Fedtschenko* (?); near Tschimirbai, *O.Fedtschenko* (?).

A.boissieri Regel, tom. cit.:75(1875). Type: Iran, near Tehran, collector not stated
(holotype ?LE).
A.lehmannianum var. *bungei* Boiss., Fl. Or. 5:235(1882). Type: Iran, near Shahrud,
Bunge (not traced).
A.stocksianum Boiss. var. *persicum* Boiss., tom. cit. 267(1882). Type: Iran, Shurab,
near Esfahan, *Bunge* 12 (not traced).

ILLUSTRATIONS. Regel, Travels in Turkestan 3,1:t.6,f.11–14(1876); Wendelbo, Dan. Biol. Skr. 10,3:172,f.61(1958), Fl. Iranica 76:t.5,f.59a,b(1971); Matine, Contr. Étude Fam. Alliaceae en Iran pl.2,no.2(1976).

Bulb oblong-ovoid, 1–1.5 cm diam.; outer tunics greyish-brown or brown, strongly reticulate-fibrous, not produced into a long neck; bulblets rather few, yellowish or pale brown, sometimes absent. Stem 10–30 cm, sometimes up to 3 from each bulb, flexuose just above ground level, more or less conspicuously ribbed, glabrous. Leaves 3–8, often exceeding the inflorescences, linear, fistulose, 1–2 mm wide, glabrous, sheathing the lower quarter to half of the stem; sheaths and lower part of the stem often strongly veined or stained purple. Spathe persistent, with 1 acuminate valve which sometimes splits at the base into 2, shorter than the umbel, silvery-membranous. Umbel hemispherical to nearly spherical, loosely-flowered, 3–5 cm diam. Pedicels unequal, 1–3 cm long, smooth, often curved upwards or suberect; bracteoles present, minute. Perianth campanulate; segments white or pinkish with a purple or sometimes green mid-vein, 5–7(–9) mm long, subequal, elliptic-ovate or elliptic-oblong, usually subacute, cucullate, smooth; Stamens included, the tip of the anthers approximately equalling the perianth segments; filaments smooth or minutely ciliate at the base, the outer ones simple, triangular-subulate, the inner ones much wider at the base than the outer, with the median cusp a quarter to a third as long as the undivided basal part and a third as long as to slightly longer than the lateral cusps; lateral cusps included within the perianth; anthers violet before dehiscence. Style not exserted. Ovary often with a corona-like stucture at the apex. Capsule subglobose, about 4 mm long. 2n=16.

FLOWERING TIME. April–June.

ECOLOGY. Gravelly or sandy desert soils and stony slopes, 900–2100m

DISTRIBUTION. C, S & E Iran; SW Pakistan(Baluchistan); Afghanistan; Turmenistan; Uzbekistan; Kazakhstan.

NOTES. Somewhat similar to *A.kotschyi* from S.Iran but with 3–8 leaves per bulb (solitary in *A.kotschyi*), a conspicuously reticulate-fibrous bulb tunic and a larger, more lax, inflorescence. Recorded as 'eaten' in Baluchistan.

101. Allium dictyoprasum *C.A.Meyer ex Kunth*, Enum. Pl. 4:390(1843), sphalm. *dyctioprasum*; Boiss., Fl. Or. 5:243(1882); Grossh., Fl. Kavk. ed.1,1:201(1928), ed.2,2:122(1940); Vved. in Fl. URSS 4:236(1935), Engl. ed. 4:183(1968); Kollmann in Davis, Fl. of Turkey 8:193(1984); Kollmann in Feinbr., Fl. Palaestina 4:94(1986). Type: Armenia, *Szovits* (holotype B[? destroyed], isotypes[?]Fl,K!,LE,P!).

?*A.multiflorum* DC. var. *violaceopurpureum* C.Koch in Linnaea 22:239(1849). Type: Turkey, Çoruh Gorge & Peterek, c.1067 m, *Koch* (holotype B[?destroyed]).

A.viride Grossh., Fl. Kavk. ed.1,1:201(1928), ed.2,2:122(1940); Vved. in Fl. URSS 4:236(1935), Engl. ed. 4:182(1968); Feinbr. in Pal. Journ. Bot., Jer. Ser. 3:16(1943); Wendelbo in Rechinger, Fl. Iranica 76:51(1971); Wendelbo in Townsend & Guest, Fl. Iraq 8:151(1985). Type: Transcaucasia, Zuvand, *Grossheim* (holotype ?TBI or ERE).

A.dictyoprasum var. *virescens* Grossh., Fl. Kavk. 1,1:201(1928). Type: Transcaucasia, no specimen cited.

A.emarginatum Rech. fil. in Ark. Bot. ser. 2,1:505(1951). Type: Lebanon, Zahle, *Mouterde* 101 (holotype S!).

ILLUSTRATIONS. Fl. URSS 4:t.14,f.1,1a(1935); Pal. Journ. Bot., Jer. Ser. 3:19,f.27(1943); Fl. Iranica 76:t.5,f.68(1971); Fl. of Turkey 8:183,f.7,no.19 & 20(1984). Plates 15B–C.

Bulb oblong-ovoid, (1.8–)2–2.5 cm diam.; outer tunics brown, reticulate-fibrous, produced into a long neck; bulblets absent or few, rather large and elongate, yellowish, with reticulate nervation. Stem 40–60(–150) cm. Leaves 3–4, shorter than the inflorescence, cylindrical, hollow, 3–11 mm wide, glabrous, sheathing to approximately the lower third of the stem; sheaths smooth. Spathe with 2, shortly beaked valves, shorter than the umbel, caducous. Umbel globose, (1–)1.5–3.5(–5) cm diam., very dense. Pedicels 0.5–1.7 cm long at anthesis; bracteoles present, numerous, white, linear. Perianth ovoid-campanulate; segments smooth, green, brownish or purplish, often with white margins, sometimes green with purple tips, 2.5–3.5 mm long, the outer ones strongly cymbiform, broadly elliptic or ovate, obtuse, rounded or emarginate, inner ones slightly longer, elliptic-obovate or elliptic-oblong, emarginate. Stamens strongly exserted, filaments glabrous or ciliate at base; outer entire or sometimes with 1–2 lateral teeth or with 3 cusps and resembling the inner ones, inner with 3–(5–7) cusps, the middle cusp shorter to slightly longer than the basal lamina and shorter than, subequal to, or slightly exceeding the lateral cusps, sometimes only half as long; all cusps well-exserted from the perianth; anthers purple or yellow. Style exserted, green, 4–4.5 mm long. Capsule subglobose, about 3.5–4 mm long; seeds black, about 2 mm long. 2n=16.

FLOWERING TIME. May–July.

ECOLOGY. Dry rocky places, mountain steppe, saline soils and volcanic craters, 100–3050 m.

DISTRIBUTION. Transcaucasia; N & W Iran; E & SE Turkey; N Iraq; W Syria; Israel; ?Jordan.

NOTES. The recognition of two taxa, *A.dictyoprasum* and *A.viride*, based on flower colour and leaf width, is unsatisfactory since there appears to be no convincing correlation between these characters and the distribution. Even within the fairly limited material available it can be seen that there is great variation in the morphology and structure of the stamens, particularly in the relative proportions of their cusps and basal lamina.

A.dictyoprasum is related to *A.karyeteinii* but has larger flowers and the outer filaments are often provided with lateral teeth or cusps; this is an unusual feature in section *Allium* and is not known to occur in *A.karyeteinii*.

A.emarginatum Rech. fil., based on *Mouterde* 101(S!) appears to belong here although in stature and leaf width it more closely resembles *A.karyeteinii*; it is probably an individual collected from a population of rather depauperate plants, perhaps during a dry season. The significant morphological characters all agree with those of *A.dictyoprasum*.

102. Allium karyeteinii *Post*, Fl. Syria 789(1896); Post, Fl. Syria, Palestine & Sinai 2:637(1933); Feinbr. in Pal. Journ. Bot., Jer. Ser. 3:17(1943); Mouterde, Nouv. Fl. Lib. et Syr. 1:270(1966); Kollmann in Davis, Fl. of Turkey 8:194(1984). Type: Syria, Syrian Desert, ravines south of Qaryetein, *G.Post* (holotype BEI). [The Keeper of the Herbarium has confirmed that the type specimen is in the Post collection].

ILLUSTRATIONS. Pal. Journ. Bot., Jer. Ser. 3:19,f.28(1943); Nouv. Fl. Lib. et Syr. 1(Atlas):t.85,f.2(1966); Fl. of Turkey 8:194,f.8,no.1(1984).

Bulb oblong, about 1.5 cm diam.; outer tunics brown, reticulate-fibrous, produced into a long neck; bulblets narrow, elongate, pale brown. Stem 20–60(–80) cm. Leaves 2–3, cylindrical, hollow, about 2 mm wide, glabrous or very slightly scabrid on the veins, sheathing the lower third to half of the stem. Spathe with 2–3 orbicular or broadly ovate, shortly beaked, valves, shorter than the umbel, persistent. Umbel globose or hemispherical, 1–3 cm diam., dense. Pedicels 3–8 mm long at anthesis; bracteoles present, white, membranous, conspicuous. Perianth campanulate, obovoid-oblong or subglobose; segments smooth, reddish or purplish with white scarious apex and sometimes also the margins, 2–2.5 mm long, oblong, obtuse or emarginate. Stamens exserted, filaments glabrous; middle cusp of inner filaments about half as long as the basal lamina and two thirds as long as the lateral cusps; all cusps exserted from the perianth; anthers purple. Capsule subglobose, about 2 mm long. 2n=16.

FLOWERING TIME. June–July.

ECOLOGY. Dry rocky places and desert, up to about 1000 m.

DISTRIBUTION. Syria, in Syrian Desert region; S.Turkey, Adana province.

NOTES. Similar to *A.dictyoprasum* but with smaller flowers, narrower leaves and often a rather shorter stature.

103. Allium fethiyense *N.Özhatay* & *B.Mathew* in Kew Bull. 51(1): in press. Type: Turkey, Muğla province, Köyceğiz to Fethiye, 54 km from Köyceğiz, roadsides, 100 m, 28 May 1988, *N.& E.Özhatay* ISTE 58655 (holotype ISTE!, isotype K!).

ILLUSTRATIONS. Kew Bull. 51(1): in press.

Bulb ovoid, about 1.5 cm diam., outer tunics brown, rather weakly reticulate-fibrous, produced at the apex into a long neck 7–15cm in length; bulblets apparently absent. Stem 40–65 cm. Leaves 3–4, shorter than the stem, terete, fistulose, 1–1.5 mm wide, glabrous, prominently ribbed, sheathing the stem for three quarters of its length. Spathe soon deciduous, broadly ovate at base, narrowed to a slender beak, 6–7 cm long [spathe details from ISTE 63014 (ISTE!)]. Umbel rather lax, hemispherical, c. 2.5 cm diam. Pedicels 7–11 mm long, subequal, smooth; bracteoles present, aristate, c. 1–1.5 mm long, membranous, white . Perianth campanulate at anthesis; segments apparently white (in dried state), membranous, 2–2.5 mm long, c. 0.5 mm wide at the base, narrowly lanceolate-oblong, obtuse or subacute, saccate at the base, strongly keeled in the lower half, smooth. Stamens with anthers well-exserted; outer filaments simple, triangular-subulate, 3.5 mm long, slightly narrower than perianth segments at the base, ciliate; inner filaments 3-cuspidate, the median cusp 1.5–2.5 times longer than the basal lamina, scarcely ciliate at the base, much shorter than the lateral cusps, all three cusps well-exserted from the perianth at anthesis; anthers yellow. Style 2 mm long. Capsule broadly ovoid-triquetrous, 2.5–3.2 mm long, longer than the perianth, valves ovate-elliptic, emarginate at apex; seeds black, 2.5–3 mm long [capsule and seeds described from ISTE 62568 (ISTE!)]. 2n=16.

FLOWERING TIME. June–July.
ECOLOGY. Stoney and rocky slopes, roadsides, under *Pinus brutia*, 80–100 m.
DISTRIBUTION. SW Turkey (recorded in Muğla province).
NOTES. In its bulb tunic characters the new species resembles *A.dictyoprasum*, although the fibres are not nearly as strongly reticulated as those of the latter. This feature, combined with the very narrow perianth segments and leaves the unusual relative proportions of the cusps of the inner filaments, make this a distinctive species.

104. Allium baytopiorum *Kollmann & N.Özhatay* in Notes R.B.G. Edinb. 41:246(1983); Kollmann in Davis, Fl. of Turkey 8:194(1984). Type: Turkey, Kars prov., foot of Agri Dağ, 1200 m, 19 May 1979, *T.Baytop* ISTE 43568 (holotype ISTE!).

ILLUSTRATIONS. Fl. of Turkey 8:194,f.8,no.2(1984).

Bulb globose, 2–3 cm diam.; outer tunics pale brown, reticulate–fibrous, produced into a long neck; bulblets few, reddish-brown. Stem 60–100 cm, thick. Leaves 3–4, cylindrical, fistulose, up to 5 mm wide, smooth, sheathing to approximately the lower half of the stem. Spathe caducous. Umbel hemispherical, 5 cm diam., dense. Pedicels 1.5–2 cm long at anthesis; bracteoles present, white, narrowly lanceolate, about 1 cm long. Perianth oblong-campanulate; segments with reflexed tips, brownish or greenish with narrow white margins, 6–6.5 mm long, the outer ones broadly elliptic, acute, minutely and densely scabrid on outer surface and margins, the inner ones oblong, truncate or emarginate, glabrous. Stamens subexserted, filaments sparsely ciliate at base; middle cusp of inner filaments shorter than the basal lamina and slightly shorter than the lateral cusps; lateral cusps about equalling the perianth; anther colour unknown. Capsule subglobose, about 4 mm long. 2n=16.

FLOWERING TIME. May
ECOLOGY. Lower mountain slopes, 1200 m.
DISTRIBUTION. Turkey: Kars province, known only from the type collection.

NOTES. Related to *A.dictyoprasum* but distinguished by its much larger perianth segments, densely scabrid on the outside.

105. Allium robertianum *Kollmann* in Notes R.B.G. Edinb. 41:264(1983); Kollmann in Davis, Fl. of Turkey 8:195(1984). Type: Turkey, Antalya prov., between Çakirli and Antalya, *Davis* 15422 (holotype E, isotype K!).

ILLUSTRATIONS. Fl. of Turkey 8:194,f.8,no.3(1984).

Bulb oblong, about 2 cm diam.; outer tunics brown, reticulate-fibrous, produced into a long neck; bulblets unknown. Stem 50–200 cm, rather thick (to 1.7 cm). Leaves withering away at or before flowering time, 2–3, cylindrical, fistulose, 0.5–1.5 cm wide, glabrous, sheathing to approximately the lower half of the stem. Spathe 1-valved, shortly beaked, caducous. Umbel globose, 3–7 cm diam., dense. Pedicels 1–3 cm long at anthesis; bracteoles present, white. Perianth cylindrical-campanulate or ellipsoid; segments smooth, whitish-green or white with a green mid-vein, 4–5 mm long, ovate-oblong, the outer acute or apiculate, the inner obtuse and sometimes apiculate. Stamens sub- exserted; middle cusp of inner filaments much shorter than the basal lamina and distinctly shorter than the lateral cusps; lateral cusps well exserted from the perianth; anthers violet. Style 3 mm long. Capsule subglobose, about 4–5 mm long; seeds black, triquetrous, with a colliculate surface, about 2.5–3 mm long.

FLOWERING TIME. May–July (–August).
ECOLOGY. Crevices of limestone cliffs and on roadsides, from sea level to 1550m.
DISTRIBUTION. Turkey: Antalya, Adana and Hatay provinces.
NOTES. According to Kollmann in Fl. of Turkey, loc. cit., this is allied to *A.hamrinense* but differs in having taller stems, broader leaves and whitish perianth segments which are longer, narrower and smooth . However, the latter has a persistent spathe with 2–3 valves, whereas that of *A.robertianum* is caducous and 1-valved, and I would regard this as an indication that they are not closely related.

106. Allium turcomanicum *Regel* in Acta Horti Petrop. 10:305(1887); Vved. in Fl. URSS 4:234(1935), Engl. ed. 4:181(1968); Vved. in Fl Uzbek. 1:451(1941); Wendelbo in Rechinger, Fl. Iranica 76:47(1971). Type: C.Asia, R.Murgab, between Kalaburan and Abdul-achan, June 1881, *A.Regel* (holotype LE, isotype K!,P!).

ILLUSTRATIONS. Acta Horti Petrop. 10:t.1,f.4(1887); Fl. Iranica 76:t.5,.61(1971).

Bulb ovoid, 1–1.5 cm diam.; outer tunics brown, reticulate-fibrous, the inner ones yellow; bulblets few, rather large, elongate, acuminate, yellow, shiny. Stem 30–100 cm, smooth, conspicuously ribbed in the dried state. Leaves much shorter than the infloresence, 4–5, semi-cylindrical, fistulose, 2–6 mm wide, minutely scabrid on the veins, sheathing the lower third to half of the stem; sheaths glabrous. Spathe 1-valved, ovate with a narrow beak, shorter than the umbel, caducous. Umbel spherical or slightly longer than wide, about 3–5.5 cm diam., fairly dense. Pedicels 1–3 cm long, with a tendency to all curve upwards, smooth; bracteoles present, small, silvery-white. Perianth campanulate-urceolate; segments whitish or pale purple-pink with a green or purplish median vein, 5–6 mm long, narrowly lanceolate, acute and apiculate, smooth, keeled. Stamens with anthers slightly exserted at anthesis; filaments ciliate at base, the outer ones simple, subulate above the broadened base, the inner ones with the anther-bearing cusp about equal in length to, or slightly longer than, the very widely expanded broadly ovate or suborbicular basal lamina and equal to or slightly longer than the lateral cusps; lateral cusps shorter than or about equal to the perianth; anthers violet. Style exserted at anthesis, about 3 mm long. Capsule subglobose or obovoid, about 3.5 mm long, furnished with conspicuous transverse wrinkles; seeds black, c. 3.5 mm long. 2n=16,32.

FLOWERING TIME. May–June.

ECOLOGY. Stony slopes, clayey or sandy soils.

DISTRIBUTION. C.Asia, Mts. of Turkmenistan and Tadjikistan; NW & NE Afghanistan.

NOTES. A distinctive species which has narrow perianth segments and very widely expanded inner filaments, the basal part wider than the adjacent segment and ovate to almost orbicular in outline.

107. Allium filidens *Regel* in Acta Horti Petrop. 3,2:174(1875); Vved. in Fl. Turkm. 1:285(1932); Vved. in Fl. URSS 4:235(1935), Engl. ed. 4:181(1968); Vved. in Fl Uzbek. 1:452(1941); Kaschtsch. & E.Nikit. in Fl. Kirgiz. 3:88(1951); Pavl. & Poljak. in Fl. Kazak. 2:183(1958); Vved. in Fl. Tadjik. 2:333(1963); Wendelbo in Rechinger, Fl. Iranica 76:48(1971). Type: C.Asia, 'in montibus Mogol-Tau', *Sewerzow* (holotype LE).

A.karakense Regel in Acta Horti Petrop. 3:76(1875). Type: C.Asia, 'in arenosis prope montes Karak, *O.Fedtschenko* (holotype LE).

A.ugami Vved. in Trans. Sc. Soc. Turk. 1:123(1923). Type: C.Asia, Syr Darya, in valley of R.Ugam, 17 June 1921, *Uranov* 895; June 1921, *Varanov* 119; 11 July 1922, *Vatueva* 173 (syntypes TAK).

ILLUSTRATIONS. Fl. Iranica 76:t.5,f.62; t.16,f.2(1971). Plate 16A.

Bulb ovoid, 1–2.5 cm diam.; outer tunics brown, reticulate-fibrous, extended into a neck at the apex; bulblets few, rather large, elongate, acuminate, yellowish-brown or dark brown. Stem 20–60 cm, sometimes apparently more than one produced per bulb, smooth, conspicuously ribbed in the dried state. Leaves shorter than the infloresence, 3–4, semi-cylindrical, canaliculate above, fistulose, 1–5(–7) mm wide, smooth or minutely scabrid on the veins, sheathing the lower fifth to a third of the stem; sheaths glabrous. Spathe 1-valved, c.3 cm long, ovate with a narrow beak slightly shorter than to about as long as the wide basal part, equalling or exceeding the umbel, caducous. Umbel spherical or hemisperical, about 3–6.5 cm diam., fairly dense. Pedicels 1–3.5 cm long, smooth; bracteoles present, small, silvery-white. Perianth campanulate-urceolate or ovoid; segments white or very pale bluish-green with a green median vein, 4–5 mm long, smooth, the outer narrowly elliptic-ovate, keeled, acute or subacute, the inner narrowly elliptic-oblong, subacute or obtuse. Stamens with anthers subequal to perianth or subexserted at anthesis; filaments ciliate at base, the outer ones simple, subulate above the broadened base, the inner ones with the anther-bearing cusp about half as long to nearly as long as the very widely expanded broadly ovate basal lamina and about half to two thirds as long as the lateral cusps; lateral cusps much-exceeding the perianth; anthers purple-brown. Style included, about 2 mm long. Capsule subglobose, about 4 mm long; seeds black, c.3.5 mm long. 2n=16.

FLOWERING TIME. May–June

ECOLOGY. Gravelly slopes, 300–2400 m.

DISTRIBUTION. C.Asia, recorded in the Aral-Caspian, Kyzyl Kum, Amu Darya, Syr

Darya, Pamir-Alai and Tien Shan regions; C,N.& E.Afghanistan, Pakistan.

NOTES. Related to *A.turcomanicum* but with the lateral cusps of the inner filaments much longer than the median cusp and long-exserted from the perianth.

108. Allium filidentiforme *Vved. ex Kaschtsch. & E.Nikit.* in Bot. Nat. Gerb. Inst. Bot. A.N. Uzbek.SSR 13:32(1952). Type: C.Asia, Tien Shan, Fergana range, 4 July 1945, *Kalinina & Moreva* 71 (holotype LE).

Bulb ovoid, about 2.5 cm diam.; outer tunics pale reddish-brown or pale brown, coriaceous, reticulate-veined, finally breaking into reticulate fibres; bulblets few, large. Stem 50–70 cm, smooth, prominently veined. Leaves 3–4, semi-cylindrical, caniculate above, fistulose, about 3 mm wide, glabrous, sheathing the lower third of the stem; sheaths glabrous. Spathe shorter than the umbel, shortly beaked, caducous. Umbel ovoid-spherical, about 5 cm diam., fairly dense. Pedicels up to 2 cm long; bracteoles present, linear-filiform, scarious. Perianth ovoid; segments 'greenish', about 4–4.5 mm long, obtuse, the outer oblong, navicular, minutely tuberculate on the keel, the inner oblanceolate, slightly shorter. Stamens with anthers exserted at anthesis; filaments ciliate at base, the outer ones simple, narrowly triangular, the inner ones with the anther-bearing cusp about equal to the basal lamina and slightly shorter than the lateral cusps; lateral cusps exserted from the perianth; anther colour unknown. Style slightly exserted at anthesis, about 2 mm long. Capsule suborbicular, 4 mm long.

FLOWERING TIME. June–July.

ECOLOGY. Mountain slopes on gypsum formations.

DISTRIBUTION. C.Asia, Fergana Mts.

NOTES. Closely related to *A.filidens* but the inner filaments have a slightly different structure, and it apparently has a different flower colour.

109. Allium margaritiferum *Vved.* in Consp. Fl. Asiae Mediae 2:316(1971). Type: C.Asia, S.Pamir-Alai, between Bajsun and Schurtschi, 1965, *Filimonova* s.n., fld. cult. 2 July 1966 Tashkent Bot. Gard., *Hort. Bot. Acad. Sci. Uz.SSR* 1002 (holotype TAK).

Bulb ovoid, 1–1.5 cm diam.; outer tunics red-brown, reticulate-fibrous; bulblets few, rather large, acuminate, yellow, smooth. Stem (30–)50–80 cm, smooth. Leaves 3–4, semi-cylindrical, caniculate above, fistulose, 1–3 mm wide, sheathing the lower quarter to third of the stem; sheaths glabrous. Spathe caducous. Umbel spherical or hemisperical, about 2–4 cm diam., fairly dense. Pedicels 1–2 cm long, smooth; bracteoles present. Perianth subglobose; segments 'pearly-lilac' with a green median vein, about 4 mm long, obtuse, the outer oblong, concave, the inner oblanceolate-oblong, slightly narrower and shorter. Stamens with anthers exserted at anthesis; filaments ciliate at base, the outer ones simple, narrowly triangular, the inner ones with the anther-bearing cusp slightly longer than the basal lamina and about equal in length to the lateral cusps; lateral cusps exserted from the perianth; anther colour unknown. Style exserted. Capsule details unknown.

FLOWERING TIME. June–July

ECOLOGY. Mountain slopes on gypsum formations.

DISTRIBUTION. C Asia, S Pamir-Alai.

NOTES. Closely allied to the other Central Asiatic species *A.brevidentiforme* and *A.filidentiforme* but differing from both in flower colour, from the former in having obtuse perianth segments and from the latter in its rather smaller umbels.

110. Allium brevidens *Vved.* in Not. Syst. Herb. Horti Petrop. 5:89(1924); Vved. in Fl. URSS 4:233(1935), Engl. ed. 4:180(1968); Vved. in Fl. Uzbek. 1:451(1941). Type: C.Asia, S. slope of Hissar range near Kara-tag, 20 May 1913, *Michelson* 1721 (syntype LE); Kuljab, 9 July 1899, *Lipsky* (syntype LE); Kuljab, Rink, *Vvedensky* 5500 (syntype LE).

ILLUSTRATIONS. Fl. URSS 4:231 (Engl. ed.179),pl.14,f.2(1935).

Bulb ovoid, 0.7–1.5 cm diam.; outer tunics brown, reticulate-fibrous, the inner ones yellow; bulblets few, helmet-shaped, pointed at both ends. Stem 20–30 cm, smooth. Leaves equalling or exceeding the infloresence, 2–3, semi-cylindrical, canaliculate above, fistulose, 1–3 mm wide, scabrous, sheathing the lower half of the stem; sheaths scabrous. Spathe 1-valved, ovate with a beak one and a half times as long as the wide basal part, caducous, slightly shorter than the umbel. Umbel spherical or subspherical, about 2.5–4 cm diam., dense. Pedicels 0.9–2 cm long at anthesis; bracteoles present. Perianth ovoid; segments whitish with a purplish median vein, 3–4 mm long, elliptic-lanceolate, acute, smooth, the outer ones keeled. Stamens with anthers exserted at anthesis; filaments glabrous, the outer ones simple, triangular-subulate, the inner ones with the anther-bearing cusp about one and a half times as long as the very widely expanded basal lamina and the lateral cusps; lateral cusps shorter than the perianth; anther colour unknown. Style well-exserted. Capsule subglobose, about 3.5 mm long. 2n=16.

FLOWERING TIME. May–July
ECOLOGY. Rocky outcrops.
DISTRIBUTION. C.Asia, Pamir-Alai mountains.
NOTES. The epithet derives from the fact that the lateral cusps of the inner filaments are shorter than the anther-bearing central cusp, an unusual, although not unique, arrangement in section *Allium*.

111. Allium brevidentiforme *Vved.* in Consp. Fl. Asiae Mediae 2:315(1971). Type: C.Asia, R.Kashkadarya basin, valley of R.Igri-su, 6 July 1955, *Pytaeva & Tsukervanik* 1617 (holotype TAK).

Bulb ovoid, 2 cm diam.; outer tunics brown, reticulate-fibrous; bulblets few, yellowish, reticulate-veined on the outside and subcrystalline. Stem 60 cm, ribbed. Leaves 4, [apparently, according to Vvedensky] semi-cylindrical, fistulose, 2–3 mm wide, sheathing the lower half of the stem. Spathe caducous. Umbel spherical, about 3–4.5 cm diam., dense. Pedicels 1.5–2 cm long; bracteoles present. Perianth ovoid; segments white with a dirty green median vein, 4–4.5 mm long, oblong, acute. Stamens with anthers slightly exserted; filaments ciliate at base, the outer ones simple, triangular-subulate, the inner ones with the anther-bearing cusp subequal in length to the expanded basal lamina and to the lateral cusps; anther colour unknown. Style well-exserted. Capsule unknown.

FLOWERING TIME. June–July.
ECOLOGY. Unknown.
DISTRIBUTION. C Asia, W Pamir-Alai mountains.
NOTES. This species is known only from the type collection and its original description, which is far from complete. It is said to differ from *A.brevidens* in having 'sub-

crystalline' bulblets and different filament structure; in *A.brevidens* the central cusp of the inner filaments is longer than the lateral cusps whereas in *A.brevidentiforme* it is subequal to them.

112. Allium valentinae *Pavl.* in Animad. Syst. Herb. Univ. Tomsk. 1-2:1(1953); Pavl. & Poljak. in Fl. Kazakh. 2:182(1958). Type: C.Asia, Dshamboul, 'montes Kurdaj, in faucibus Karapetak', 28 June 1952, *V.P.Michailova* (holotype AA, isotype MHA,TAK).

Bulb ovoid, 1–2 cm diam.; outer tunics brown, reticulate-fibrous; bulblets few, fairly large, yellowish, smooth. Stem 40–50 cm, smooth, ribbed. Leaves 2–3, semi-terete, fistulose, smooth, shorter than the inflorescence, sheathing the lower quarter to third of the stem; sheaths smooth. Spathe lanceolate, shorter than the umbel, caducous. Umbel spherical, up to 6 cm diam., dense. Pedicels 2–3 cm long; bracteoles present. Perianth ovoid; segments pale yellowish with a darker brownish median vein, 4–4.5 mm long, narrowly lanceolate, acuminate, the outer slightly shorter, keeled. Stamens with anthers just exserted at anthesis; outer filaments simple, triangular, the inner ones with the anther-bearing cusp about a third as long as the basal lamina and about equal in length to the lateral cusps; lateral cusps about equalling or slightly exserted from the perianth; anthers greyish. Style included, about 2 mm long. Capsule subglobose, 3.5–4 mm long.

FLOWERING TIME. May–June.
ECOLOGY. Mountain slopes.
DISTRIBUTION. C.Asia, N.Tien Shan Mts., Mt. Kurdaj.
NOTES. According to Vvedensky, related to *A.filidentiforme* but with yellowish flowers with acuminate segments and filaments slightly shorter than the perianth.

113. Allium crystallinum *Vved.* in Bull. Univ. As. Centr. 19:126(1934); Vved. in Fl. URSS 4:234(1935), Engl. ed. 4:181(1968); Vved. in Fl. Uzbek. 1:452(1941); Vved. in Fl. Tadj. 2:333(1963). Type: C.Asia, Pamir-Alai, Ketmenchapta range near Aulata, 22 May 1930, *Lepeshkin* (holotype TAK).

Bulb ovoid, about 2 cm diam.; outer tunics grey-brown, subcoriaceous, reticulate-fibrous, the outer ones with shiny (originally described as 'crystalline') protuberances, the inner ones yellow; bulblets few, rather large, yellow. Stem about 60 cm, smooth. Leaves shorter than the infloresence, 2, cylindrical, fistulose, 3–5 mm wide, smooth, sheathing the lower third of the stem; sheaths smooth. Spathe 1-valved, ovate with a narrow beak, slightly exceeding the umbel, caducous. Umbel subspherical, 2.5–3.5 cm diam., fairly dense. Pedicels 1–1.5 cm long, smooth; bracteoles present. Perianth ovoid-campanulate; segments whitish with a reddish median vein, 5 mm long, acute, smooth, the outer ones oblong–lanceolate, keeled, the inner ones linear-lanceolate and slightly longer. Stamens with anthers included or equalling the perianth; filaments ciliate at base, the outer ones simple, triangular-subulate, the inner ones with the anther-bearing cusp half as long as the very widely expanded basal lamina and half as long as the lateral cusps; lateral cusps about equal to the perianth; anther colour unknown. Style included. Capsule unknown. 2n=16,32.

FLOWERING TIME. May
ECOLOGY. Stony slopes, among junipers.
DISTRIBUTION. C.Asia, Pamir-Alai Mts.
NOTES. Known only from the type collection; it is apparently related to *A.turco-*

manicum but the structure of the inner filaments is rather different.

114. Allium sosnowskyanum *Miscz.* ex *Grossh.*, Fl.Kavk. ed.1,1:207(1928); Kollmann in Davis, Fl. of Turkey 8:195(1984). Type: Turkey, Erzurum province, Oltu, 18 June 1911, *Sosnowsky* (lectotype TBI!).

ILLUSTRATIONS. Fl. Kavk. ed.2,2:t.15,f.1(1940); Fl. of Turkey 8:194,f.8,no.4(1984). Plates 16B–C.

Bulb ovoid, 1–1.5 cm diam., outer tunics dark brown, reticulate-fibrous; bulblets ?not produced. Stem 10–20 cm. Leaves exceeding the stem, 2(–3), flattish, fistulose, 2–3 mm wide, glabrous, sheathing the lower third to half of the stem; sheaths slightly inflated, distinctly winged-ribbed, the wings somewhat undulate. Spathe caducous, rather small. Umbel globose, 2.5–4 cm diam. Pedicels 1–1.8 cm long. Perianth broadly campanulate; segments white or pale pink with a purple or green mid-vein, 4–5 mm long, broadly ovate, acute, scabrid outside and slightly denticulate on margin. Stamens included; median cusp of inner filaments about one third as long as the basal lamina and the lateral cusps, the lateral cusps slightly exserted from the perianth. Capsule subglobose, about 4 mm long. 2n=16.

FLOWERING TIME. June–July.

ECOLOGY. Clearings in *Picea orientalis* forest and *Quercus petraea* scrub, and on dryish rocky slopes, 1270–1720m.

DISTRIBUTION. Turkey: Çoruh, Erzurum and Gümüşane provinces.

NOTES. A very distinctive species, unlikely to be confused with any other with its low stature, cylindrical leaves longer than the scape and white flowers with a dark median vein.

Species which may be confused with Section Allium

115. Allium turkestanicum *Regel* in Acta Horti Petrop. 3:197(1875); op. cit. 10:351(1887); Vved. in Fl. Turkm. 1:281(1932); Vved. in Fl. URSS 4:229(1935), Engl. ed. 4:177(1968); Vved. in Fl. Uzbek. 1:449(1941); Kaschtsch. & E.Nikit. in Fl. Kirgiz. 3:87(1951); Pavl. & Poljak. in Fl. Kazakh. 2:181(1958); Wendelbo in Rechinger, Fl. Iranica 76:50(1971); Tzagolova in Bot. Mater. Gerb. Inst. Bot. Akad. Nauk Kazakh. SSR 11:44(1979). Type: C.Asia, 'in montibus Mogol-Tau', *Kuschakewicz* (holotype LE).

A.nothum Vved. in Opred. rast. okr. Tashkent. 1:66(1923), nom. nud.

ILLUSTRATIONS. Regel, Reise in Turkestan 3,1:t.15,f.6-8(1876); Fl. Iranica 76:t.5,f.65(1971). Plate 16D.

Bulb ovoid, 1.5–3 cm diam.; outer tunics greyish or sometimes purplish, subcoriaceous; bulblets absent. Stem (40–)50–100 cm, smooth. Leaves 4–6, linear, flat, non-fistulose, 8–10 mm wide, scabrid-papillose on the margins, shorter than the inflorescence, sheathing the lower quarter of the stem, withering away before flowering time; sheaths smooth or scabrid. Spathe 2-valved, the valves with a beak, equalling the umbel, subcoriaceous and prominently ribbed, persistent. Umbel spherical, 1.5–4 cm diam., dense. Pedicels 0.7–2 cm long; bracteoles present, linear, scarious. Flowers fragrant (sweet with a hint of onions!). Perianth campanulate; segments pale pink with a raised median vein which is almost the same colour as the rest of the lamina, 3–4 mm long, 1–1.5 mm wide, obtuse, the outer narrowly elliptic-lanceolate, the inner narrowly elliptic-oblong. Stamens with anthers exserted at anthesis; outer filaments simple, triangular-subulate, the inner ones with the anther-bearing cusp about three quarters as long as, or subequal to, the elliptic or obovate basal lamina and about six times longer than the very short tooth-like obtuse lateral cusps; lateral cusps much shorter than the perianth; anthers purplish-lilac, becoming brownish in the dried state. Style exserted at anthesis. Capsule subglobose, 3.5 mm long, with a transverse-rugose surface; seeds black, about 3 mm long. 2n=16.

FLOWERING TIME. June–July.

ECOLOGY. Sandy, clayey & rocky slopes.

DISTRIBUTION. C Asia, Kopet Dag & Tien Shan ranges, Balkhash region.

NOTES. *A.turkestanicum* should probably be treated as a member of sect. *Codonoprasum*. Vvedensky excluded it from sect. *Allium*, presumably because the inner filaments are merely toothed rather than provided with long filiform appendages. Tzagolova op. cit. (1979) described a new series to accommodate it within the former section and Khassanov (1992) also regards it as belonging to *Codonoprasum*.

I have included it in the key to the species of section *Allium* since its overall appearance is similar to that of some species in this Section, and the presence of distinct teeth on the filaments might well lead to some confusion.

Other species not in Section *Allium*, but having small teeth on the inner filaments, are mentioned on page 53.

SELECTED BIBLIOGRAPHY

Astley, D., Innes, N.L. & Meer, Q.P. van der (1982). Genetic Resources of *Allium* species. IBPGR, Rome.

Borg, J. (1927). *Allium*. Descr. Fl. of the Maltese Islands 683.

Boscher, J. & Auger, J. (1991). L'*Allium ampeloprasum* (var. *bulbilliferum* Lloyd) de l'Île d'Yeu (Vendée) est chimiquement un ail et non un poireau. Bull. Soc. Bot. France Lett. Bot. 138(4–5):315–320.

Bothmer, R. von (1970). Cytological studies in *Allium*. Chromosome numbers and morphology in sect. *Allium* from Greece. Bot. Notiser. 123:519–551.

––––– (1972). Four species of *Allium* sect. *Allium* in Greece. Bot. Notiser 125:62–76.

––––– (1974). Biosystematic studies in the *Allium ampeloprasum* complex. Opera Bot. (Lund) 34:1–104.

––––– (1975). The *Allium ampeloprasum* complex on Crete. Mitt. Bot. Staatssamml. München 12:267–288.

––––– (1982). Karyotype variation in *Allium commutatum*. Plant Syst. Evol. 140:179–182.

Don, G. (1827). A monograph of the genus *Allium*. Edinburgh, an advance reprint of Mem. Wernerian Nat. Hist. Soc. 6:1–102(1832).

Etoh, T., Kojima, T., & Matsuzoe, N. (1992). Fertile garlic clones collected in Caucasia, in Hanelt, P., Hammer, K. & Knupffer (eds.), The genus *Allium* — Taxonomic Problems and Genetic Resources 49–54.

Feinbrun, N. (1943). *Allium* sect. *Porrum* of Palestine and the neighbouring countries. Palestine Journ. Bot., Jer. Ser. 3(1):1–21.

––––– (1948). Further studies on *Allium* of Palestine and the neighbouring countries. Palestine Journ. Bot., Jer. Ser. 4:144–157.

Friesen, N.V.(1987). *Allium*, in Malyschev, L.I. & Peschkova, G.A., Flora Sibiriae, Araceae-Orchidaceae 55–96, 177–195.

Garbari, F. & Cela Renzoni, G. (1975). Il genere *Allium* in Italia, 7. Il caso di *Allium commutatum* Guss. Lavori Soc. Ital. Biogeogr. N.S. 5:1–16.

Guern, M., Le Corff, J. & Boscher, J. (1991). Caryologie comparée des *Allium* du groupe *ampeloprasum* en France. Bull. Soc. Bot. France Lett. Bot. 138(4–5):303–313.

Hanelt, P., Hammer, K. & Knupffer (eds.), The genus *Allium* — Taxonomic Problems and Genetic Resources: Proceedings of an International Symposium held at Gatersleben, Germany, June 11–13, 1991:107–123.

Hanelt, P. et al. (1992). Infrageneric Groupings of *Allium* — The Gatersleben Approach, in Hanelt, P., Hammer, K. & Knupffer (eds.), The genus *Allium* — Taxonomic Problems and Genetic Resources 107–123.

Helm, J. (1956). Die zu Wurz– und Speisezwecken kultivierten Arten der Gattung *Allium*. Kulturpflanze 4:130–180.

Hermann, F. (1939). Sectiones et subsectiones nonnullae Europaeae generis *Allium*. Feddes Repert. 46:421–422.

Janka, V. de (1885). Amaryllideae, Dioscoreae et Liliaceae Europeae analytice elaboratae. Termés. Fuz. 90:41–77; pp. 50–60 transl. into English by W.T.Stearn as 'Key to the Alliums of Europe' in Herbertia 11:219–225(1946).

Jones, H.A. & Mann, L.K. (1963). Onions and their Allies. Interscience Publishers, New York.

Kazakova, A.A. (1978). *Allium* in Zhukovsky, P.M.(Ed.), Fl. of Cultivated Plants 10. Kolos, Leningrad.

Khassanov, F.O. (1992). A revision of the genus *Allium* L. in the flora of Uzbekistan, in Hanelt, P., Hammer, K. & Knupffer (eds.), The genus *Allium* — Taxonomic

Problems and Genetic Resources 153–159.

Kollmann, F. (1971). *Allium ampeloprasum* L. in Israel. Israel Journ. Bot. 20:263–272.

―――― (1984). *Allium* L. in Davis, P.H.(Ed.), Fl. of Turkey 8:98–211. University Press, Edinburgh.

―――― (1986). *Allium* L. in Feinbrun–Dothan, N.(Ed.), Fl. Palaestina 4(text):74–99,413–414. Israel Academy of Sciences, Jerusalem.

―――― & Shmida, A. (1977). *Allium* species of Mt. Hermon. Israel Journ. Bot. 26:128–148.

Maire, R. (1958). *Allium* in Fl. de l'Afrique du Nord 5:244–304.

Matine, F. (1976). Contribution a l'étude de la Famille Alliaceae en Iran. Institut de Recherches Entomologiques et Phytopathologiques d'Evine, Teheran. Dept. de Botanique No. 6.

Meikle, R.D. (1985). *Allium* L. in Fl. of Cyprus 2:1608–1628.

Moore, H.E. (1955). The cultivated Alliums IV. Baileya 3:156–167.

Mouterde, P. (1966). Nouvelle Fl. du Liban et de la Syrie 1(texte):257–285.

Nasir, E.(1975). Alliaceae, in Nasir, E. & Ali, S.I., Flora of West Pakistan 83:1–31.

Omelczuk, T. (1962). Sistematicheckii sklad. Ukraini rid. *Allium* L. Ukrain. Bot. Zur. 19(3):66–73.

Özhatay, N. (1984). Cytotaxonomic studies of the genus *Allium* in European Turkey and around Istanbul. 3, sect. *Allium* and sect. *Melanocrommyum*. Journ. Fac. of Pharmacy Istanbul 20:43–65.

Ohwi, J. (1965). *Allium*, in Flora of Japan: 294–296.

Pastor, J. & Valdes, B. (1983). Revision del genero *Allium* (Liliaceae) en la Peninsula Iberica e Islas Baleares. Universidad de Sevilla, Seville.

Peters, R.J., Stolte, A. & Van der Werff, M. Selection for increased rate of vegetative multiplication in a wild collection of *Allium ampeloprasum*. Eucarpia, Proceedings of the 4th *Allium* Symposium 1988, Wellesbourne.

Pogosian, A.I.(1983). Chromosome numbers of some species of the genus *Allium* (Alliaceae) distributed in Armenia and Iran. Bot. Zhurn. 68,5:652(1983).

Regel, E. (1875). Alliorum adhuc cognitorum Monographia. Acta Horti Petrop. 3(2)1–266.

Richter, K. (1890). Plantae Europeae 1:198–210.

Shmida, A. & Kollmann, F. (1977). *Allium* species of Mt. Hermon. Israel Journ. Bot. 26:149–159.

Stearn, W.T. (1960). *Allium* and *Milula* in the central and eastern Himalaya. Bull. Brit. Mus. Nat. Hist. Bot. 2:159–191

―――― (1978). European species of *Allium* and allied genera of Alliaceae: a synonymic enumeration. Ann. Mus. Goulandris 4:83–198.

―――― (1980). *Allium* L. in Tutin, T.G. et al. (Eds.) Fl. Europaea 5:49–69.

―――― (1981). The genus *Allium* in the Balkan Peninsula. Bot. Jahrb. 102:201–213.

―――― & Campbell, E. (1986). *Allium* in Walters, S.M. et al.(Eds.). European Garden Flora 1:233–246.

―――― & Özhatay, N. (1977). *Allium sphaerocephalon* and an allied species of European Turkey, *A.proponticum*. Ann. Mus. Goulandris 3:45–50.

Täckholm, V. & Drar, M. (1954). Flora of Egypt 3:58–136.

Traub, H.P. (1968). The subgenera, sections and subsections of *Allium* L. Plant Life 24:147–163.

―――― (1972). Genus *Allium* L.; subgenera, sections and subsections. Plant Life 28:132–137.

Tscholokoschvili, N. (1975). Ad cognitionem systematis generis *Allium* L. Notulae Syst. Georg. Inst. Bot. Thbiliss. 31:36-54.

_____ (1977). Conspectus systematis spocierum Caucasicarum generis *Allium* L. Notulae Syst. Georg. Inst. Bot. Thbiliss. 34:21–33.

Tzagolova, V.G. (1975). Notulae Systematicae De Sectiones Porrum G.Don Generere *Allium* L. Kazachstania. Bot. Mater. Gerb. Inst. Bot. Akad. Nauk Kaz. S.S.R. 9:10

_____ (1977). Materies ad Systema Sectionis *Molium* in Kazachstania. Bot. Mater. Gerb. Inst. Bot. Akad. Nauk Kaz.S.S.R. 10:13.

_____ (1979). Ad Systema Generis Alii Sectionis *Codonoprasi* in Kazachstania Notula. Bot. Mater. Gerb. Inst. Bot. Akad. Nauk Kaz.S.S.R. 11:43.

Vvedensky, A.I. (1935). *Allium* in Fl. URSS 4:112–280. Translated into English by Airy–Shaw, H.K. as 'The genus *Allium* in the USSR'. Herbertia 11:65–218; also English edition Fl. USSR 4:87–216. Jerusalem (1968).

Wendelbo, P. (1971). Alliaceae in Rechinger, K.H. Fl. Iranica 76:1–100.

_____ (1985). *Allium* L. in Townsend, C.C. & Guest, E.(Eds.) Fl. of Iraq 8:137–180.

Wilde–Duyfjes, B.E.E. de (1973). Typification of 23 *Allium* species described by Linnaeus and possibly occurring in Africa. Taxon 22(1):57–91.

_____ (1976). A revision of the genus *Allium* L. (Liliaceae) in Africa, Proefschrift. Wageningen. Also published as Meded. Landbouwhogesch. Wageningen 76:no.11.

Xu, J.M. (1980). *Allium*, in Flora Reip. Pop. Sinicae 14:170–272.

Zangheri, P.(1976). *Allium*, in Flora Italica 1:861.

Note added in proof (see p.135)

A. Soldano has recently pointed out the existence of the combination *A. margaritaceum* subsp. *tenorei* (Parl.) A. Terracc. in Malpighia 3: 299 (1889) (as 'tenorii'). this is based on *A. margaritaceum* var. *tenorei* Parl. which is considered synonymous with *A. guttatum* subsp. *sardoum*; the correct name and citation for this taxon thus appears to be *A. guttatum* subsp. *tenorei* (Parl.) Soldano in Thaiszia 4: 120 (1994)

APPENDIX 1: *Allium* sect. *Allium*

LIVING MATERIAL STUDIED

(species in alphabetical order)

Notes:

The first set of brackets show the collector and collector's number.

ISTE = International herbarium code for Istanbul University, Faculty of Pharmacy.

It.An. = Iter Anatolicum, an expedition to Turkey in 1990 comprising N. & E. Özhatay (Istanbul University), J. Cowley, M. Johnson & M.J. Doherty (Kew); and F. Garbari & A. Giordani (Pisa University).

The figure in square brackets is the Kew accession number.

Cult.BM = cultivated by B. Mathew.

The last set of figures (where shown) are the Kew cytological reference numbers and their associated chromosome counts. Published chromosome counts have author/s and year following the cytological data. Where DNA estimations have been made, (unpublished), these are indicated [DNA]. * New chromosome data. See Table 1, pp. 22–31.

[82] *A. affine* (It.An.36-134) Turkey [1990-2112] 90-234, 2n=16

[82] *A. affine* (It.An.49-205) Turkey [1990-2155] 90-277, 2n=16 [DNA]

    A. affine (It.An.51-211) Turkey [1990-2158] 90-279, 2n=16

    A. affine (It.An.52-219) Turkey [1990-2163] 90-284, 2n=16

    A. affine (M.Koyuncu 8754) Turkey [1990-2461] 90-398, 2n=16 [DNA]*

[33] *A. albiflorum* (Gabrielian s.n.) N.Caucasus [cult.BM]

[31] *A. alibile* (Collenette 7201) Saudi Arabia [1989-1569]

[81] *A. amethystinum* (Horton/Stevens 968) Yugoslavia [1975-4275]

    A. amethystinum (N. Özhatay ISTE 60507) Turkey [1989-2748]

    A. amethystinum (N. Özhatay s.n.) Turkey [1976-40409] 76-100 & 90-3, 2n=16

[1] *A. ampeloprasum* var. *babingtonii* (Kruger s.n.) England [1973-1883] 75-218, 2n=48 (Johnson 1982)

    A. ampeloprasum var. *babingtonii* (Robinson s.n.) Ireland [1977-2898]

    A. ampeloprasum (Baines/Henry 108b) Turkey [cult.BM] 91-59, 2n=24*

    A. ampeloprasum (Cowley 2) Corsica [1989-1736] 89-888, 2n=32*

    A. ampeloprasum (Cowley 3) Corsica [1989-1737] 89-889, 2n=32*

    A. ampeloprasum (Ferguson 4110A) Majorca [1985-8594] 90-4, 2n=32*

    A. ampeloprasum (Gray s.n.) Italy [1987-0792]

    A. ampeloprasum (It.An.23-99) Turkey [1990-2089] 90-214, 2n=32 [DNA]

    A. ampeloprasum (It.An.47-195) Turkey [1990-2150] 90-272, 2n=32

    A. ampeloprasum (It.An.78-353) Turkey [1990-2219] 90-341, 2n=32 [DNA]

    A. ampeloprasum (Kammerlander/Pasche 90/105) Turkey [cult.BM]

    A. ampeloprasum (Kruger s.n.) England [1973-1884]

    A. ampeloprasum (Kruger s.n.) England [1973-1885]

    A. ampeloprasum (Kruger s.n.) England [1973-1886]

    A. ampeloprasum (Norman 1190) Spain [1990-1075]

    A. ampeloprasum (N. Özhatay ISTE 32063) Turkey [1976-4018] 76-102, 2n=40

    A. ampeloprasum (N. Özhatay ISTE 30130) Turkey [1976-4020] 76-105, 2n=40

    A. ampeloprasum (N. Özhatay s.n.) Turkey [1976-4035]

    A. ampeloprasum (N. Özhatay ISTE 57059) Turkey [1986-3647]

    A. ampeloprasum (Roderick s.n.) Turkey [cult.BM]

     A. ampeloprasum (Stearn s.n.) Greece [1977-1733] 77-1723, 2n=32 (Johnson 1982)

[91] *A. artemisietorum* (Fragman s.n.) Israel [1990-1171]
     A. artemisietorum (Fragman s.n.) Israel [1990-1342]

[74] *A. artvinense* (Baines/Henry 76) Turkey [cult.BM]
     A. artvinense (Grabrielian s.n.) Armenia [cult.BM]

[42] *A. asperiflorum* (It.An.30-117) Turkey [1990-2098] 90-373, 2n=16

[13] *A. atroviolaceum* (Baines/Henry 116b) Turkey [cult.BM]
     A. atroviolaceum (Baines/Henry 117) Turkey [1989-2200] 90-380, 2n=32
     A. atroviolaceum (Baines/Henry 59) Turkey [1989-2143]
     A. atroviolaceum (Collenette) Saudi Arabia [1987-4043]
     A. atroviolaceum (Furse) Iran [1962-7146] 90-7, 2n=32
     A. atroviolaceum (It.An.43-180) Turkey [1990-2143] 90-265, 2n=24 [DNA]
     A. atroviolaceum (It.An.12-50) Turkey [1990-2060]
     A. atroviolaceum (It.An.27-105) Turkey [1990-2091] 90-216, 2n=32
     A. atroviolaceum (It.An.35-132) Turkey [1990-2110] 90-232, 2n=32 [DNA]
     A. atroviolaceum (It.An.37-143) Turkey [1990-2118] 90-240, 2n=32 [DNA]
     A. atroviolaceum (It.An.39-163) Turkey [1990-2133] 90-255, 2n=16
     A. atroviolaceum (It.An.45-184) Turkey [1990-2144] 90-266, 2n=24 [DNA]
     A. atroviolaceum (Johnson 306) Turkey [1985-2275]
     A. atroviolaceum (N. Özhatay ISTE 57058) Turkey [1986-3651]
     A. atroviolaceum (N. Özhatay ISTE 32105) Turkey [1976-4052] 76-97, 2n=16
     A. troviolaceum (Sayers s.n.) Iran [1968-0380] 90-6, 2n=32*
     A. atroviolaceum (Sayers s.n.) Iran[1968-0403] 90-10, 2n=32*
     A. atroviolaceum (Sharman 15002) Turkey [1989-2750] 89-958, 2n=32
     A. atroviolaceum (Sharman 15005) Turkey [1989-2753]
     A. atroviolaceum (Sharman 15012) Turkey [1989-2760]

[58] *A. aucheri* (Koyuncu 8696) Turkey [1990-2459] 90-354, 2n=16 [DNA]
     A. aucheri (Sharman 15008) Turkey [cult.BM] 90-94, 2n=16

[3c] *A. borgeaui* ssp. *creticum* (Turland) Crete [cult.BM]

[100] *A. borszczowii* (Fritsch KAZ.955193) Uzbekistan [cult.BM]

[15] *A. cappadocicum* (It.An.18-73) Turkey [1990-2073]
     A. cappadocicum (It.An.6-27) Turkey [1990-2049]
     A. cappadocicum (Koyuncu 8567) Turkey [1990-2478] 90-375, 2n=16*
     A. cappadocicum (Koyuncu 8795) Turkey [1990-2464] 90-399, 2n=16 [DNA]*
     A. cappadocicum (N. Özhatay ISTE 60411) Turkey [1989-2746] 89-954, 2n=16 [DNA]

[84] *A. chamaespathum* (Johnson 540) Greece [1989-3204] 89-1023, 2n=16*

[2]   *A. commutatum* (Cowley 5) Corsica [1989-1739] 89-891, 2n=16*
     A. commutatum (Cowley s.n.) Paxos (Greece) [1980-1729] 90-13, 2n=16 [DNA]*
     A. commutatum (Gray s.n.) Italy [1987-0793]
     A. commutatum (Kuhbier s.n.) Balearic Is. [1980-2735] 90-16, 2n=24 [DNA]*
     A. commutatum (Sands 3293) Greece [1978-2387] 90-15, 2n=16 [DNA]*
     A. commutatum (Sands 3294) Greece [1978-2388] 78-977, 2n=16*
     A. commutatum (Sands 3295) Greece [1978-2385]
     A. commutatum (Sands 3456) Greece [1980-1725] 90-12, 2n=16 [DNA]*
     A. commutatum (Stearn s.n.) Greece [1977-1734] 77-1724, 2n=16 (Johnson 1982) [DNA]

[113] *A. crystallinum* (Fritsch UZB.850193) Uzbekistan [cult.BM]

[67] *A. curtum* (Ben Gurian Univ. s.n.) Israel [1984-0153]

       A. curtum (Ben Gurian Univ. s.n.) Israel [1984-0154] 90-17, 2n=32*
       A. curtum (It.An.55-231) Turkey [1990-2167] 90-288, 2n=16
       A. curtum (It.An.69-317) Turkey [1990-2202] 90-232, 2n=16;16+4B
       A. curtum (Salmon/Lovell 37) Jordan [cult.BM]
[101] *A. dictyoprasum* (Baines/Henry 53) Turkey [cult.BM] 90-365, 2n=16*
       A. dictyoprasum (collector not known) Iran [1989-1951]
       A. dictyoprasum (Gabrielian s.n.) Armenia [cult.BM]
       A. dictyoprasum (It.An.28-107) Turkey [1990-2093] 90-218, 2n=16
       A. dictyoprasum (It.An.57-233) Turkey [1990-2168] 90-289, 2n=16
       A. dictyoprasum (Koyuncu 8787) Turkey [1990-2465] 90-397, 2n=16
       A. dictyoprasum (Koyuncu 8811) Turkey [1990-2467] 90-406, 2n=16
       A. dictyoprasum (Markus 673) Turkey [1989-2247] 89-904, 2n=16;32
       A. dictyoprasum (N. Özhatay ISTE 60471) Turkey [1989-2745]
       A. dictyoprasum (Stevens 13) Turkey [1990-1585]
       A. dictyoprasum (Stevens 32) Turkey [1990-1603]
       A. dictyoprasum (Stevens 33)Turkey [1990-1604]
       A. dictyoprasum (Stevens 66)Turkey [1990-1636]
[47] *A. dregeanum* (Perry s.n.) S.Africa [1990-0801]
[46] *A. enginii* (It.An.71-323) Turkey [1990-2204] 90-325, 2n=16
[44] *A. erubescens* (Furse 7127) Iran [cult.BM] 90-81, 2n=16 [DNA]*
       A. erubescens (Gabrielian s.n.) Armenia [cult.BM]
[107] *A. filidens* (Stevens s.n.) Uzbekistan [cult.BM]
[66] *A. fuscoviolaceum* (Baines/Henry 116c) Turkey [cult.BM] 90-89, 2n=16*
       A. fuscoviolaceum (It.An.12-51) Turkey [1990-2061]
       A. fuscoviolaceum (Johnson 306) Turkey [1985-2275]
       A. fuscoviolaceum (Koyuncu 8738) Turkey [1990-2468] 90-409, 2n=16 [DNA]
       A. fuscoviolaceum (Marais 1577A) Turkey [1976-6059]
       A. fuscoviolaceum (Sharman 15004) Turkey [1989-2752]
       A. fuscoviolaceum (Sharman 15006) Turkey [1989-2754]
       A. fuscoviolaceum (Stevens 18) Turkey [1990-1589]
       A. fuscoviolaceum (Stevens 31) Turkey [1990-1602]
       A. fuscoviolaceum (Vasak s.n.) Georgia 1989-2257]
[52] *A. gomphrenoides* (Archibald 5131A) Greece [1984-3642] 90-19 & 90-410,
       2n=16*
[25] *A. gramineum* (Baines/Henry 141) Turkey [1989-2183] 90-366, 2n=16
[80a] *A. guttatum* ssp. *guttatum* (Archibald 5226) Turkey [1984-8233]
       A. guttatum ssp. *guttatum* (Baines 16) Turkey [cult.BM] 90-369, 2n=16*
       A. guttatum ssp. *guttatum* (It.An.1-3) Turkey [1990-2034] 90-158, 2n=16
       A. guttatum ssp. *guttatum* (Mathew 1015) Turkey [1985-2301]
       A. guttatum ssp. *guttatum* (Mericli/Sutlupinar ISTE 56954A) Turkey [1986-
       3644] 86-817,2n=32
       A. guttatum ssp. *guttatum* (Mericli/Sutlupinar ISTE 56948)Turkey[1986-3638]
       A. guttatum ssp. *guttatum* (N. Özhatay ISTE 30527)Turkey[1976-4025]
       A. guttatum ssp. *guttatum* (N.Özhatay ISTE 58828)Turkey[1989-2732]
       A. guttatum ssp. *guttatum* (Pasche 78/19)Turkey[1978-3178]
       A. guttatum ssp. *guttatum* (Stevens 78)Turkey[1990-1648]
       A. guttatum ssp. *guttatum* (N. Özhatay ISTE 58828) Turkey [1989-2732]
[80b] *A. guttatum* ssp. *sardoum* (Baxter s.n.) Corfu [1974-3476]
       A. guttatum ssp. *sardoum* (Cook/Keesing 15) Greece [1978-4763] 90-23,
       2n=16*
       A. guttatum ssp. *sardoum* (Cook/Keesing 21) Greece [1978-4769] 90-25,
       2n=16*

A. *guttatum* ssp. *sardoum* (Hoog/Paul 859) Greece [1988-4156] 90-24, 2n=16*
A. *guttatum* ssp. *sardoum* (Johnson 548) Greece [1989-3211] 89-1013, 2n=32*
A. *guttatum* ssp. *sardoum* (Norman 190) Spain [1990-1066]
A. *guttatum* ssp. *sardoum* (Norman 490) Spain [1990-1069]
A. *guttatum* ssp. *sardoum* (Schilling 3024) Greece [1990-0095]
A. *guttatum* ssp. *sardoum* (Stearn s.n.) Greece [1976-1269] 78-959, 2n=16*
A. *guttatum* ssp. *sardoum* (Stearn s.n.) Greece [1977-1735] 90-22&78-37, 2n=16*

[55] A. *heldreichii* (Meikle s.n.) Greece [1969-0804] 71-425, 2n=16 (Johnson 1982)
       A. *heldreichii* (Trevan/Whitehead 6) Greece [1971-4170]
[88] A. *hierochuntinum* (Salmon/Lovell 79) Jordan [1988-2784] 90-80, 2n=16*
[5]   A. *iranicum* (Fliegner/Simmons 477) Iran [1977-4268]
[57] A. *jubatum* (It.An.2-4) Turkey [1990-2035] 90-159, 2n=16
       A. *jubatum* (Mertens/Pasche/Richter 79-42) Turkey [1989-2261]
[54] A. *junceum* (Georgiades 10) Cyprus [1987-1345]
       A. *junceum* (Hewer 4763) Cyprus [1981-2205]
       A. *junceum* (Johnson 514) Turkey [1988-2745]
[54b] A. *junceum* ssp. *tridentatum* (N. Özhatay ISTE 60504) Turkey [1989-2743]
[102] A. *karyeteinii* (It.An.41-178) Turkey [1990-2141] 90-263, 2n=16
       A. *karyeteinii* (Koyuncu 8758) Turkey [1990-2471] 90-402, 2n=16 [DNA]*
[1b] A. *kurrat* (seed from L.Boulos) Egypt [cult.BM]
[6]   A. *leucanthum* (?source) [1969-9525]
[19] A. *longicuspis* (I.P.K.Gatersleben, Germany) C.Asia [cult.BM]
       A. *longicuspis* (Ruksans s.n.) C.Asia [cult.BM]
[14] A. *macrochaetum* (It.An.40-170) Turkey [1990-2137] 90-259, 2n=16
       A. *macrochaetum* (Koyuncu 8768A) Turkey [1990-2462] 90-408, 2n=16 [DNA]*
       A. *macrochaetum* (N. Ö zhatay ISTE 60462) Turkey [1989-2744] 89-952, 2n=16
       A. *macrochaetum* (Stevens 51) Turkey [1990-1622]
[59] A. *nevsehirense* (Baines/Henry 52) Turkey [cult.BM] 90-362, 2n=32
       A. *nevsehirense* (It.An.18-74) Turkey [1990-2074] 90-200, 2n=16 [DNA]
       A. *nevsehirense* (It.An.3-9) Turkey [1990-2040] 90-164, 2n=16 [DNA]
       A. *nevsehirense* (Koyuncu 8799) Turkey [1990-2474] 90-407, 2n=16 [DNA]
       A. *nevsehirense* (N. Özhatay ISTE 60410) Turkey [1990-2742]
[22] A. *oltense* (Baines/Henry 116d) Turkey [cult.BM]
       A. *oltense* (Johnson 248) Turkey [1985-2209]
       A. *oltense* (Koyuncu 8747) Turkey [1990-2475] 90-355, 2n=16 [DNA]*
       A. *oltense* (Koyuncu 8810) Turkey [1990-2472] 90-400, 2n=16 [DNA]*
[60] A. *phanerantherum* (Fragman s.n.) Israel [1990-1172]
[1a] A. *porrum* (several cvs.) [cult.BM; cult.HRI Wellesbourne] 90-429, 2n=32 [DNA]*
[64a] A. *proponticum* var. *proponticum* (N. Özhatay ISTE 33050) Turkey[1976-4028] 76-106, 2n=16
[4]   A. *pseudoampeloprasum* (Baines/Henry 108c) Turkey [cult.BM]
       A. *pseudoampeloprasum* (Baines/Henry 54) Turkey [cult.BM]
[28] A. *pustulosum* (It.An.31-123) Turkey [1990-2103]
[12] A. *pyrenaicum* (Guyancourt B.G., via Gatersleben)France[cult.K]
[23] A. *rollovii* (Sharman 15015)Turkey[1989-2763]
[41b] A. *rotundum* ssp. *jajlae* (Baines/Henry 11) Turkey [1989-2142] 90-363, 2n=32
       A. *rotundum* ssp. *jajlae* (Stevens 67) Turkey [1990-1637]
       A. *rotundum* ssp. *jajlae* (Suckow s.n.) Crimea [cult.BM]

[41a] *A. rotundum* ssp. *rotundum* (Baines/Henry 125) Turkey [1989-2147]

A. rotundum ssp. *rotundum* (Baines/Henry 22) Turkey [1989-2199]

A. rotundum ssp. *rotundum* [Koyuncu 8750] Turkey [1990-2457] 90-393, 2n=16 [DNA]*

A. rotundum ssp. *rotundum* (Koyuncu 8776) Turkey [1990-2458] 90-370, 2n=16 & 32*

A. rotundum ssp. *rotundum* (Koyuncu 8804) Turkey [1990-2463] 90-401, 2n=32*

A. rotundum ssp. *rotundum* (It.An.48-201) Turkey [1990-2152] 90-274, 2n=16 [DNA]

A. rotundum ssp. *rotundum* (Sharman 15010) Turkey [1989-2758]

A. rotundum ssp. *rotundum* (Stevens 63) Turkey [1990-1634]

A. rotundum ssp. *rotundum* (Stevens 64) Turkey [1990-1635]

A. rotundum ssp. *rotundum* (Stevens 73a) Turkey [1990-1643]

A. rotundum ssp. *rotundum* (Stevens 74B) Turkey [1992-1428] 92-25, 2n=16*

A. rotundum ssp. *rotundum* (Stevens 76) Turkey [1990-1646]

A. rotundum ssp. *rotundum* (Stevens s.n.) Turkey [1989-8259]

A. rotundum ssp. *rotundum* (Vasak s.n.) Georgia [1989-2265]

A. rotundum ssp. *rotundum* (Archibald 5226) Turkey [1984-8086]

A. rotundum ssp. *rotundum* (Baines 18) Turkey [cult.BM] 90-87, 2n=40

A. rotundum ssp. *rotundum* (Baines 19) Turkey [cult.BM] 90-367, 2n=32

A. rotundum ssp. *rotundum* (Baines/Henry 108a) Turkey [cult.BM]

A. rotundum ssp. *rotundum* (Baines/Henry 116a) Turkey [cult.BM]

A. rotundum ssp. *rotundum* (Baines/Henry 55) Turkey [cult.BM]

A. rotundum ssp. *rotundum* (Baines/Henry 57) Turkey [cult.BM]

A. rotundum ssp. *rotundum* (Fragman s.n.) Israel [1990-1173]

A. rotundum ssp. *rotundum* (Haritonidou s.n.) Greece [1978-2008]

A. rotundum ssp. *rotundum* (It.An.7-31) Turkey [1990-2053]

A. rotundum ssp. *rotundum* (It.An.12-47) Turkey [1990-2058] 90-184, 2n=16 [DNA]

A. rotundum ssp. *rotundum* (It.An.15-63) Turkey [1990-2068] 90-194, 2n=16+2B [DNA]

A. rotundum ssp. *rotundum* (It.An.19-78) Turkey [1990-2078] 90-204, 2n=32 [DNA]

A. rotundum ssp. *rotundum* (It.An.31-121) Turkey [1990-2101] 90-223, 2n=32 [DNA]

A. rotundum ssp. *rotundum* (It.An.35-133) Turkey [1990-2111]

A. rotundum ssp. *rotundum* (It.An.42-179) Turkey [1990-2142]

A. rotundum ssp. *rotundum* (It.An.58-239) Turkey [1990-2170]

A. rotundum ssp. *rotundum* (Kerndorff/Pasche 90-07) Turkey [cult.BM]

A. rotundum ssp. *rotundum* (Koyuncu 8559) Turkey [1990-2476] 90-389, 2n=32*

A. rotundum ssp. *rotundum* (Koyuncu 8760) Turkey [1990-2469] 90-391, 2n=32 [DNA]*

A. rotundum ssp. *rotundum* (Koyuncu 8793) Turkey [1990-2477] 90-390, 2n=32*

A. rotundum ssp. *rotundum* (Koyuncu 8808) Turkey [1990-2460] 90-376, 2n=32*

A. rotundum ssp. *rotundum* (Marais 1581) Turkey [1976-1405]

A. rotundum ssp. *rotundum* (Marais 1595) Turkey [1976-1414]

A. rotundum ssp. *rotundum* (Melville/Wrigley s.n.) Yugoslavia [1965-52209]

A. rotundum ssp. *rotundum* (Norman 590) Spain [1990-1070]

A. rotundum ssp. *rotundum* (N.Özhatay s.n.) Turkey [1976-4017] 76-95, 2n=32

A. rotundum ssp. *rotundum* (N.Özhatay s.n.) Turkey [1976-4055]

A. rotundum ssp. *rotundum* (N.Özhatay/Sariyer ISTE 57084) Turkey [1986-3648]

A. rotundum ssp. *rotundum* (RHS Lily Group seed list) USSR [1984-498] 90-52, 2n=16*

A. rotundum ssp. *rotundum* (Salmon s.n.) Morocco [1989-1882] 89-585, 2n=32*

A. rotundum ssp. *rotundum* (Sharman 15003) Turkey [1989-2751]

A. rotundum ssp. *rotundum* (Sharman 15009) Turkey [1989-2757]

A. rotundum ssp. *rotundum* (Sharman 15011) Turkey [1989-2759]

A. rotundum ssp. *rotundum* (Sharman 15016) Turkey [1989-2764] 89-972, 2n=32*

A. rotundum ssp. *rotundum* (Snderhousen 1237) Turkey [cult.BM]

A. rotundum ssp. *rotundum* (Snderhousen 1256) Turkey [cult.BM]

A. rotundum ssp. *rotundum* (Stearn s.n.) Greece [1977-1742] 77-1722, 2n=32 (Johnson 1982)*

A. rotundum ssp. *rotundum* (Stevens 10) Turkey [1990-1582]

A. rotundum ssp. *rotundum* (White 4/90) Turkey [cult.BM]

[41c] *A. rotundum* ssp. *waldsteinii* (Baines/Henry 140) Turkey [1989-2145] 90-377, 2n=24

A. rotundum ssp. *waldsteinii* (Mathew 10804) Turkey [1985-2297]

[53] *A. rubrovittatum* (Barclay 132) Crete [1969-18833]

A. rubrovittatum (Barclay 160) Crete [1969-51217]

A. rubrovittatum (Barclay 292) Crete [1972-10301]

A. rubrovittatum (Barclay 52) Crete [1973-14201]

[17] *A. sandrasicum* (N.Özhatay ISTE 60503) Turkey [1989-2741] 89-949, 2n=16 [DNA]

A. sandrasicum (N.Hzhatay ISTE 60506) Turkey [1989-2738] 89-946, 2n=16

A. sandrasicum (N.Ozhatay ISTE 60508) Turkey [1989-2740] 89-948, 2n=16

[19a] *A. sativum* var. *ophioscorodon* (Suthering s.n.) [cult.K,1993]

A. sativum (several commercial stocks) [cult.BM]

[89] *A. scabriflorum* (Archibald 6079) Turkey [1985-5497] 90-5, 2n=16*

A. scabriflorum (It.An.78-352A) Turkey [1990-2217] 90-339, 2n=16

A. scabriflorum (It.An.78-352B) Turkey [1990-2218] 90-340, 2n=16

A. scabriflorum (Koyuncu 8768) Turkey [1992-1599] 90-403, 2n=16 [DNA]*

A. scabriflorum (Koyuncu 8768) Turkey [1990-2473] 90-396, 2n=16* [DNA]

A. scabriflorum (N.Özhatay ISTE 59214) Turkey [1989-2727]

[14] *A. sosnowskyanum* (Baines/Henry 93) Turkey [cult.BM]

A. sosnowskyanum (Johnson 89)Turkey[1982-2982)

A. sosnowskyanum (Sønderhousen 1255) Turkey [cult.BM]

A. sosnowskyanum (White 1/90) Turkey [cult.BM]

A. sosnowskyanum (White 2/90) Turkey [cult.BM]

[62] *A. sphaerocephalon* (Archibald 4808) Greece [1984-0865]

A. sphaerocephalon (Archibald 4809A) Greece [1984-8133] 90-42, 2n=16*

A. sphaerocephalon (Bowles 13) Greece [1975-2926] 78-976 & 90-59,2 n=16*

A. sphaerocephalon (Bowles 1) Greece [1975-2915] 78-974, 2n=16*

A. sphaerocephalon (Cox s.n.) Spain [1977-3385]

A. sphaerocephalon (Fliegner/Howard 113) Spain [1980-3117] 90-64, 2n=16*

A. sphaerocephalon (Fliegner/Howard 186A) Spain [1980-6203] 90-60, 2n=16*

A. sphaerocephalon (Fliegner/Howard 32) Spain [1980-3043]

A. sphaerocephalon (Fliegner/Howard 3) Spain [1980-3019] 90-62, 2n=32*
A. sphaerocephalon (Fliegner/Howard 94) Spain [1980-3100] 90-49, 2n=16*
A. sphaerocephalon (Halliwell 1310) Spain [1975-5384] 75-1359, 2n=16+0-1B*
A. sphaerocephalon (Halliwell 1380) Spain [1975-5426] 75-1358, 2n=16*
A. sphaerocephalon (Halliwell 1498) Spain [1975-5510] 75-1345, 2n=16*
A. sphaerocephalon (It.An.19-79) Turkey [1990-2079] 90-205, 2n=32 [DNA]
A. sphaerocephalon (It.An.33-128) Turkey [1990-2107] 90-229, 2n=16
A. sphaerocephalon (It.An.81-368) Turkey [1990-2229] 90-350, 2n=16 [DNA]
A. sphaerocephalon (N. Özhatay ISTE 30458) Turkey [1976-4037] 76-92, 2n=16
A. sphaerocephalon (N. Özhatay ISTE 58815) Turkey [1989-2733]
A. sphaerocephalon (N. Özhatay ISTE 33770) Turkey [1976-4060]
A. sphaerocephalon (Stearn s.n.) Greece [1977-1743] 77-1721, 2n=16 (Johnson 1982)
[62c] *A. sphaerocephalon* ssp. *arvense* (Marr 2737) Italy [1972-3047] 90-44, 2n=16*
A. sphaerocephalon ssp. *arvense* (It.An.45-185) Turkey[1990-2146]
[62a] *A. sphaerocephalon* ssp. *sphaerocephalon* (Norman 1490) Spain [1990-1079]
A. sphaerocephalon ssp. *sphaerocephalon* (Norman 1590) Spain [1990-1080]
A. sphaerocephalon ssp. *sphaerocephalon* (Norman 890) Spain [1990-1073]
A. sphaerocephalon ssp. *sphaerocephalon* (Norman 990) Spain [1990-1074]
[87] *A. stearnianum* (Johnson 178) Turkey [1985-2138] 90-56, 2n=16
A. stearnianum (Johnson 90) Turkey [1982-2983] 90-55, 2n=16
[87a] *A. stearnianum* ssp. *stearnianum* (Stevens 83) Turkey [1990-1653]
[71] *A. stylosum* (It.An.16-69) Turkey [1990-2071] 90-197, 2n=16
A. stylosum (Koyuncu 8560) Turkey [1990-2479] 90-392, 2n=16 [DNA]*
[26] *A. trachycoleum* (It.An.4-17) Turkey [1990-2043] 90-167, 2n=32
A. trachycoleum (N.Özhatay ISTE 60451) Turkey [1989-2739] 89-947, 2n=48
A. trachycoleum (Stevens 24) Turkey [1990-1595]
A. trachycoleum (Stevens 75) Turkey [1990-1645]
A. trachycoleum (Stevens 80) Turkey [1990-1650]
[18] *A. truncatum* (Fragman s.n.) Israel [1990-1174]
[20] *A. tuncelianum* (Johnson 128) Turkey [1982-3021] 90-51, 2n=16 [DNA]
A. tuncelianum (N.Özhatay ISTE 57067) Turkey [1986-3639]
A. tuncelianum (N. Özhatay ISTE 57101A) Turkey [1986-3643]
A. tuncelianum (N. Özhatay ISTE 57138) Turkey [1986-3636]
[115] *A. turkestanicum* (Fritsch 781) Uzbekistan [1992-1478]
[85] *A. vineale* (Baines/Henry 91) Turkey [cult.BM] 90-86, 2n=32
A.vineale (Brenan 9925) England [1972-0727] 90-57, 2n=32;40*
A. vineale (Cowley 4)Corsica[1989-1739]89-890, 2n=16*
A. vineale (Halliwell/Mason/Smallcombe 1292) Spain [1975-5375]
A. vineale (It.An.14/90-62) Turkey [1990-2066] 90-193, 2n=32 [DNA]
A. vineale (Mathew s.n.) France[cult.BM]
A. vineale (Norman 9002) Portugal [cult.BM]
A. vineale (N. Özhatay ISTE 32058) Turkey[1976-4022] 76-454, 2n=32
A. vineale (N. Özhatay s.n.) Turkey [1976-4048]
A. vineale (Salmon s.n.) Morocco [1989-1879]
A. vineale (Sharman 15007) Turkey [1989-2755]
A. vineale (Stevens 12) Turkey [1990-1584]
A. vineale (Stevens 26) Turkey [1990-1597]
A. vineale (Stevens 27) Turkey [1990-1615]
A. vineale (Stevens 44) Turkey [1990-1598]

A. vineale (Stevens 50) Turkey [1990-1621]
A. vineale (Stevens 52) Turkey [1990-1623]
A. vineale (Stevens 79) Turkey [1990-1649]
A. vineale (Stevens 81) Turkey [1990-1651]
[110] *A.* sp. aff. *brevidens* (Vasak s.n.) Kirgizia [1989-2252]

INDEX TO EPITHETS

Those in roman type are accepted taxa; synonyms are in italics; main description pages in bold. The list of accepted names (pp.49-51) and the key (pp.53-66) have not been indexed.